THE ESSENTIAL **JOHN NASH**

THE ESSENTIAL

JOHN NASH

EDITED BY

HAROLD W. KUHN AND SYLVIA NASAR

PRINCETON UNIVERSITY PRESS   PRINCETON AND OXFORD

Published by Princeton University Press, 41 William Street, Princeton, New Jersey 08054

In the United Kingdom: Princeton University Press, 3 Market Place, Woodstock, Oxfordshire OX20 1SY

ISBN 0-691-09527-2

Library of Congress Control Number: 2001095709

British Library Cataloging-in-Publication Data is available

This book has been composed in Adobe Garamond and Gill Sans.
Printed on acid-free paper.∞

www.pup.princeton.edu

Printed in the United States of America

10 9 8 7 6 5 4 3 2

CONTENTS

PREFACE

HAROLD W. KUHN

I have known John Nash for more than fifty years. We were graduate students together in the late 1940s. And, although he went off to MIT, we have never wholly lost touch. Even today, we have offices in the Mathematics Department of Princeton University on the same floor of Fine Hall. As his friend and colleague, it has been a great pleasure to co-edit this volume with his biographer, Sylvia Nasar.

In Nasar's splendid introduction to this volume, we are taken on a guided tour of Nash's scientific life, starting with his brilliant early career, the decades of mental illness that followed, and the subsequent transformation of his life by the award of the Nobel prize in economics in 1994. For me, the defining moment that divided the period when he was in the depths of his despair from his reentry into a world that he had always deserved took place on a bench in front of a minimalist Japanese fountain at the Institute for Advanced Studies in Princeton in October 1994.

Nash and I had just attended a seminar given by Herbert S. Wilf on the generation of identities for hypergeometric functions. I had called Nash earlier to ask him to lunch after the seminar but had not revealed my real purpose. Several months before, I had learned that it was almost certain that Nash would share the Nobel in an award that would recognize the central importance of non-cooperative games in modern economic theory. Using a variety of subterfuges, photos had been taken, a curriculum vitae had been assembled, an appointment to a nominal research position at Princeton had been arranged, and various supporting materials had been sent to Stockholm. In addition, with the active support of Jackie Savani, the press officer of Princeton University, preparations had been made for a press conference on the following day. Three days earlier, I had been informed that the Social Science Class of the Swedish Academy had approved the award unanimously, and I had been given permission to tell Nash the great news. I had already told Alicia Nash to take the day off from work and sworn her to secrecy.

So, as we sat on the bench, enjoying the mild fall weather and the splendor of the Institute woods, I told John that he should be up at 6:30 A.M. the following morning to receive a phone call from Carl-Olof Jacobsen, secretary general of the Nobel Foundation, who would tell him that he was sharing the Prize in Economic Sciences in Memory of Alfred Nobel. John took the news very calmly; it appeared that his son, John David Stier, had sent him an article from the *Boston Globe* saying that he was in contention for the prize and that the only impediment was the fear that his mental illness might lead to behavior that would embarrass the King of Sweden. He seemed more interested in the fact that the prize was split three ways and, after taxes, the net amount would not be that much.

We then went home to lunch where, after many objections, Nash met Jackie Savani. Nash refused to discuss any possible questions that might be asked at the press conference the next day. The press conference went very well, largely due to John's highly developed sense of humor, which turned aside questions that probed his private life with quiet, always logical, answers. On the morning of the announcement, he avoided reporters by coming, at my invitation, to my undergraduate course, where we were starting a section on game theory. It was a morning that those students will long remember.

Recognition is a cure for many ills; although Nash's mental illness had faded into remission in the years preceding 1994, the announcement of the Nobel prize signaled a new period in his life. The monetary amount of the prize was nontrivial (but subject to taxes in the United States, which is a surprise to most American Nobelists). Much more important was the recognition that was so long in coming.

Although he is one of the most original mathematical minds of the twentieth century, the reasons for this delay are easy to understand. His formally published work consists of about fifteen papers, five in game theory and ten in pure mathematics, produced in the main during the ten-year period from 1949 to 1959. In the past forty years, he has published very little and many people who valued his early work thought he was dead. The truth, as we well know, is that he was suffering from a debilitating mental illness diagnosed as schizophrenia and was living a quiet and secluded life near the academic community of Princeton, sheltered physically and emotionally by his wife, Alicia Nash.

After more than twenty-five years of life isolated in the main from academic activities, Nash began to emerge from the shadows of his mental illness and a number of friends and colleagues began to provide for him the rewards that he deserved. In 1993, at the instigation of Peter Sarnak and Louis Nirenberg, I joined them to collect and edit a sort of "Festschrift," consisting of papers in the fields in which Nash had made early and important contributions. Nash's name had been put forward for many years by a number of economists who are asked to make nominations for the Nobel Memorial Prize in Economics. In 1994, their efforts bore fruit. After the Nobel prize, honors appeared from all sides. He was elected to the National Academy of Sciences; the citation reads: "Nash is best known for his work in game theory. He has also made basic and extremely important contributions which have profoundly influenced differential geometry, real algebraic geometry, and partial differential equations." He was awarded the Steele Prize for a Seminal Contribution of the American Mathematical Society; the citation reads, in part: "This is one of the great achievements in mathematical analysis in this century." He has received honorary degrees from Carnegie Mellon University and the University of Athens, and many awards from mental-health organizations that wish to recognize his exemplary recovery.

In preparing this volume, Sylvia Nasar and I wish to enlarge the extent of this recognition by making the most important contributions of

HAROLD W. KUHN

Preface

John Nash—both in game theory and in pure mathematics—available to a wider audience. We believe that, in this form, we may also bridge the gap between the economists, on the one hand, and the pure mathematicians, on the other, each of whom has appreciated only part of Nash's scientific contributions. This book is not a "Complete Works"; we both wish John Nash well in his current research and hope to see many more significant works from him in the future.

x   Acknowledgments

Sylvia Nasar and I thank the wonderful and wonderfully congenial team at Princeton University Press for their patience, skill, and, above all, infectious enthusiasm. In particular, we thank Linny Schenck, for her careful shepherding of the production process; Tracy Baldwin, for giving the book a very special appearance; Grady Klein, for giving us an inspired book cover; and Vickie Kearn and Peter Dougherty, for their calm stewardship throughout. All editors should be so lucky!

We are also grateful to John and Alicia Nash, Martha Legg, and John Stier for their contributions, and to Avinash Dixit for his many useful suggestions. Working with old friends and making new ones was, for us, the best part of what has been a most rewarding experience.

INTRODUCTION

SYLVIA NASAR

When Freeman Dyson, the physicist, greeted John Forbes Nash, Jr. at the Institute for Advanced Study one day in the early 1990s, he hardly expected a response. A mathematics legend in his twenties, Nash had suffered for decades from a devastating mental illness. A mute, ghost-like figure who scrawled mysterious messages on blackboards and occupied himself with numerological calculations, he was known around Princeton only as "the Phantom."

To Dyson's astonishment, Nash replied. He'd seen Dyson's daughter, an authority on computers, on the news, he said. "It was beautiful," recalled Dyson. "Slowly, he just somehow woke up."

Nash's miraculous emergence from an illness long considered a life sentence was neither the first, nor last, surprise twist in an extraordinary life.

The eccentric West Virginian with the movie star looks and Olympian manner burst onto the mathematical scene in 1948. A one-line

In describing John Nash's contributions to economics and mathematics, I drew from essays by Avinash Dixit, John Milnor, Roger Myerson, and Ariel Rubinstein as well as from my biography, *A Beautiful Mind.* Avinash Dixit and Harold Kuhn kindly commented on my draft. Any errors are, of course, mine alone.

letter of recommendation—"This man is a genius."—introduced the twenty-year-old to Princeton's elite math department. A little more than a year later, Nash had written the twenty-seven-page thesis that would one day win him a Nobel.

Over the next decade, his stunning achievements and flamboyant behavior made Nash a celebrity in the mathematics world. Donald Newman, a mathematician who knew him in the early 1950s, called him "a bad boy, but a great one." Lloyd Shapley, a fellow graduate student at Princeton, said of Nash, "What redeemed him was a clear, logical, beautiful mind."

Obsessed with originality, disdainful of authority, supremely self-confident, Nash rushed in where more conventional minds refused to tread. "Everyone else would climb a peak by looking for a path somewhere on the mountain," recalled Newman. "Nash would climb another mountain altogether and from that distant peak would shine a searchlight back on the first peak."

By his thirtieth birthday, Nash seemed to have it all: he was married to a gorgeous young physicist and was about to be promoted to full professor at MIT; *Fortune* magazine had just named him one of the brightest stars of the younger generation of "new" mathematicians.

Less than a year later, however, the brilliant career was shattered. Diagnosed with paranoid schizophrenia, Nash abruptly resigned from MIT and fled to Paris on a quixotic quest to become a world citizen. For the next decade, he was in and out of mental hospitals. By forty, he'd lost everything: friends, family, profession. Only the compassion of his wife, Alicia, saved him from homelessness. Sheltered by Alicia and protected by a handful of loyal former colleagues, Nash haunted the Princeton campus, in the thrall of a delusion that he was "a religious figure of great, but secret importance."

While Nash was lost in his dreams, his name surfaced more and more often in journals and textbooks in fields as far-flung as economics and biology, mathematics and political science: "Nash equilibrium," "Nash bargaining solution," "Nash program," "De Georgi-Nash," "Nash embedding," "Nash-Moser theorem," "Nash blowing up."

Outside Princeton, scholars who built on his work often assumed he was dead. But his ideas were very much alive, becoming more influential even as their author sank deeper into obscurity. Nash's contributions to pure mathematics—embedding of Riemannian manifolds, existence of solutions of parabolic and elliptic partial differential equations—paved the way for important new developments. By the 1980s, his early work in game theory had permeated economics and helped create new fields within the discipline, including experimental economics. Philosophers, biologists, and political scientists adopted his insights.

The growing impact of his ideas was not limited to the groves of academe. Advised by game theorists, governments around the world be- gan to auction "public" goods from oil drilling rights to radio spectra, reorganize markets for electricity, and devise systems for matching doctors and hospitals. In business schools, game theory was becoming a staple of management training.

The contrast between the influential ideas and the bleak reality of Nash's existence was extreme. The usual honors passed him by. He wasn't affiliated with a university. He had virtually no income. A small band of contemporaries had always recognized the importance of his work. By the late 1980s, their ranks were swelled by younger scholars who launched a fight to get Nash long-overdue recognition. They succeeded spectacularly: in 1994, after an explosive behind-the-scenes debate and a narrow vote, the Swedish Academy of Sciences awarded Nash a Nobel prize in economics for his early work on non-cooperative games. The prize, which he shared with Reinhard Selten and John Harsanyi, was more than an intellectual triumph; it was a victory for those who believed that mental illness shouldn't be a bar to the ne plus ultra of scientific honors.

Most Nobel laureates, while celebrated within their disciplines, remain invisible to the public at large. And a Nobel rarely changes winners' lives profoundly. Nash is an exception. "We helped lift him into daylight," said Assar Lindbeck, chairman of the Nobel prize committee. "We resurrected him in a way."

Recognition of his ideas has not only redeemed the man—bringing him back to society and mathematics—but has turned Nash into something of a cultural hero. Since winning the Nobel, the mathematician who spent his life "thinking, always thinking" has inspired a *New York Times* profile; a biography, *A Beautiful Mind*; a *Vanity Fair* article; a

Broadway play, *Proof*; and, now, a Hollywood movie, directed by Ron Howard and starring Russell Crowe as Nash.

The ongoing celebration of Nash's inspiring life and unique achievements has generated new interest in the seminal papers he published during his twenties. *The Essential John Nash* was conceived to make these articles accessible to a wide audience. This volume reflects the full range of Nash's diverse contributions. For the first time, readers will have the opportunity to see for themselves why Nash, so nearly forgotten, has been called "the most remarkable mathematician of the second half of the century."*

Nash arrived in Princeton on the first day of Truman's 1948 re-election campaign and found himself suddenly at the center of the mathematical universe. The demigods of twentieth century science were in residence: Einstein, Gödel, Oppenheimer, and John von Neumann. "The air is full of mathematical ideas and formulae," one of Einstein's assistants marveled. It was a heady time. "The notion was that the human mind could accomplish anything with mathematical ideas," one of Nash's fellow graduate students recalled.

The ten or so first-year students were a cocky bunch, but Nash was even cockier. He loved sparring in the common room. He avoided classes. He was rarely seen cracking a book. Pacing endlessly, whistling Bach, he worked inside his own head. John Milnor, the topologist, who was a freshman that year, said, "It was as if he wanted to rediscover, for himself, three hundred years of mathematics." Always on the lookout for a shortcut to fame, Nash would corner visiting lecturers, clipboard and writing pad in hand. "He was very much aware of unsolved problems," said Milnor. "He really cross-examined people."

He was bursting with ideas. Norman Steenrod, Nash's faculty adviser, recalled:

During his first year of graduate work, he presented me with a characterization of a simple closed curve in the plane. This was essentially the same one given by Wilder in 1932. Some time later

---

* Mikhail Gromov, 1997.

he devised a system of axioms for topology based on the primitive concept of connectedness. I was able to refer him to papers by Wallace. During his second year, he showed me a definition of a new kind of homology group which proved to be the same as the Reidemeister group based on homotopy chains.

Nash's first mathematical coup, appropriately enough, involved a game of his own invention. One afternoon von Neumann strolled into the common room to see two students hunched over an unfamiliar game board. Oh, by the way, what was it that they were playing? he later asked a colleague. "Nash," came the answer, "Nash."

Parker Bros. later called Nash's nifty game, which was invented independently by the Danish mathematician Piet Hein, Hex. Nash's playful foray into mathematical games foreshadowed a far more serious involvement in a novel branch of mathematics (see chapter 3, this volume).

Today, the language of game theory permeates the social sciences. In 1948, game theory was brand-new and very much in the air at Princeton's Fine Hall. The notion that games could be used to analyze strategic thinking has a long history. Such games as Kriegspiel, a form of blind chess, were used to train Prussian officers. And mathematicians like Emile Borel, Ernst Zermelo, and Hugo Steinhaus studied parlor games to derive novel mathematical insights. The first formal attempt to create a theory of games was von Neumann's 1928 article, "Zur Theorie der Gesellschaftsspiele," in which he developed the concept of strategic interdependence. But game theory as a basic paradigm for studying decision making in situations where one actor's best options depend on what others do did not come into its own until World War II when the British navy used the theory to improve its hit rate in the campaign against German submarines. Social scientists discovered it in 1944 when von Neumann and the Princeton economist Oskar Morgenstern published their masterpiece, *Theory of Games and Economic Behavior*, in which the authors predicted that game theory would eventually do for the study of markets what calculus had done for physics in Newton's day.

The pure mathematicians around the university and the Institute were inclined to view game theory as "just the latest fad" and "déclassé"

Introduction

because it was applied, not pure mathematics. But in the eyes of Nash and his fellow graduate students, von Neumann's interest in the field lent it instant glamor.

Nash wrote his first major paper—his now-classic article on bargaining—while attending Albert Tucker's weekly game theory seminar during his first year at Princeton, where he met von Neumann and Morgenstern. But he had come up with the basic idea as an undergraduate at Carnegie Tech—in the only economics course (international trade) he ever took.

Bargaining is an old problem in economics. Despite the rise of the marketplace with millions of buyers and sellers who never interact directly, one-on-one deals—between individuals, corporations, governments, or unions—still loom large in everyday economic life. Yet, before Nash, economists assumed that the outcome of a two-way bargaining was determined by psychology and was therefore outside the realm of economics. They had no formal framework for thinking about how parties to a bargain would interact or how they would split the pie.

Obviously, each participant in a negotiation expects to benefit more by cooperating than by acting alone. Equally obviously, the terms of the deal depend on the bargaining power of each. Beyond this, economists had little to add. No one had discovered principles by which to winnow unique predictions from a large number of potential outcomes. Little if any progress had been made since Edgeworth conceded, in 1881, "The general answer is . . . contract without competition is indeterminate."

In their opus, von Neumann and Morgenstern had suggested that "a real understanding" of bargaining lay in defining bilateral exchange as a "game of strategy." But they, too, had come up empty. It is easy to see why: real-life negotiators have an overwhelming number of potential strategies to choose from—what offers to make, when to make them, what information, threats, or promises to communicate, and so on.

Nash took a novel tack: he simply finessed the process. He visualized a deal as the outcome of either a process of negotiation or else independent strategizing by individuals each pursuing his own interest. Instead of defining a solution directly, he asked what reasonable conditions any division of gains from a bargain would have to satisfy. He then posited four conditions and, using an ingenious mathematical argument, showed that, if the axioms held, a unique solution existed that maximized the product of the participants' utilities. Essentially, he reasoned, how gains are divided reflects how much the deal is worth to each party and what other alternatives each has.

By formulating the bargaining problem simply and precisely, Nash showed that a unique solution exists for a large class of such problems. His approach has become the standard way of modeling the outcomes of negotiations in a huge theoretical literature spanning many fields, including labor-management bargaining and international trade agreements.

Since 1950, the Nash equilibrium—Nash's Nobel-prize-winning idea—has become "*the* analytical structure for studying all situations of conflict and cooperation."*

Nash made his breakthrough at the start of his second year at Princeton, describing it to fellow graduate student David Gale. The latter immediately insisted Nash "plant a flag" by submitting the result as a note to the *Proceedings of the National Academy of Sciences*. In the note, "Equilibrium Points in n-Person Games," Nash gives the general definition of equilibrium for a large class of games and provides a proof using the Kakutani fixed point theorem to establish that equilibria in randomized strategies must exist for any finite normal form game (see chapter 5).

After wrangling for months with Tucker, his thesis adviser, Nash provided an elegantly concise doctoral dissertation which contained another proof, using the Brouwer fixed point theorem (see chapter 6). In his thesis, "Non-Cooperative Games," Nash drew the all-important distinction between non-cooperative and cooperative games, namely between games where players act on their own "without collaboration or communication with any of the others," and ones where players have opportunities to share information, make deals, and join coalitions. Nash's theory of games—especially his notion of equilibrium for such games (now known as Nash equilibrium)—significantly extended the boundaries of economics as a discipline.

* Roger Myerson 1999.

SYLVIA NASAR

Introduction

All social, political, and economic theory is about interaction among individuals, each of whom pursues his own objectives (whether altruistic or selfish). Before Nash, economics had only one way of formally describing how economic agents interact, namely, the impersonal market. Classical economists like Adam Smith assumed that each participant regarded the market price beyond his control and simply decided how much to buy or sell. By some means—i.e., Smith's famous Invisible Hand—a price emerged that brought overall supply and demand into balance.

Even in economics, the market paradigm sheds little light on less impersonal forms of interaction between individuals with greater ability to influence outcomes. For example, even in markets with vast numbers of buyers and sellers, individuals have information that others do not, and decide how much to reveal or conceal and how to interpret information revealed by others. And in sociology, anthropology, and political science, the market as explanatory mechanism was even more inadequate. A new paradigm was needed to analyze a wide array of strategic interactions and to predict their results.

Nash's solution concept for games with many players provided that alternative. Economists usually assume that each individual will act to maximize his or her own objective. The concept of the Nash equilibrium, as Roger Myerson has pointed out, is essentially the most general formulation of that assumption. Nash formally defined equilibrium of a non-cooperative game to be "a configuration of strategies, such that no player acting on his own can change his strategy to achieve a better outcome for himself." The outcome of such a game must be a Nash equilibrium if it is to conform to the assumption of rational individual behavior. That is, if the predicted behavior doesn't satisfy the condition for Nash equilibrium, then there must be at least one individual who could achieve a better outcome if she were simply made aware of her own best interests.

In one sense, Nash made game theory relevant to economics by freeing it from the constraints of von Neumann and Morgenstern's two-person, zero-sum theory. By the time he was writing his thesis, even the strategists at RAND had come to doubt that nuclear warfare, much less post-war reconstruction, could usefully be modeled as a game in which the enemy's loss was a pure gain for the other side. Nash had the critical insight that most social interactions involve neither pure competition nor pure cooperation but rather a mix of both.

From a perspective of half a century later, Nash did much more than that. After Nash, the calculus of rational choice could be applied to situations beyond the market itself to analyze the system of incentives created by any social institution. Myerson's eloquent assessment of Nash's influence on economics is worth quoting at length:

> Before Nash, price theory was the one general methodology available to economics. The power of price theory enabled economists to serve as highly valued guides in practical policy making to a degree that was not approached by scholars in any other social science. But even within the traditional scope of economics, price theory has serious limits. Bargaining situations where individuals have different information . . . the internal organization of a firm . . . the defects of a command economy . . . crime and corruption that undermine property rights. . . .
>
> The broader analytical perspective of non-cooperative game theory has liberated practical economic analysis from these methodological restrictions. Methodological limitations no longer deter us from considering market and non-market systems on an equal footing, and from recognizing the essential interconnections between economic, social, and political institutions in economic development. . . .
>
> By accepting non-cooperative game theory as a core analytical methodology alongside price theory, economic analysis has returned to the breadth of vision that characterized the ancient Greek social philosophers who gave economics its name.*

Von Neumann, however, didn't think much of Nash's breakthrough. When Nash met with him, the Hungarian polymath dismissed the younger man's result as "trivial." The 1953 edition of his and Morgenstern's *Theory of Games and Economic Behavior* included only a perfunctory mention of "non-cooperative games" in the Preface.

---

* Myerson 1999.

SYLVIA NASAR

Introduction

His doctorate in his pocket, Nash headed off to RAND, the ultra-secret cold war think tank, in the summer of 1950. He would be part of "the Air Force's big-brain-buying venture"—whose stars would eventually serve as models for Dr. Strangelove—for the next four years, spending every other summer in Santa Monica.

Game theory was considered RAND's secret weapon in a nuclear war of wits against the Soviet Union. "We hope [the theory of games] will work, just as we hoped in 1942 that the atomic bomb would work," a Pentagon official told *Fortune* at the time. Nash got an excited reception. Researchers like Kenneth Arrow, who won a Nobel for his social choice theory, were already chafing at RAND's "preoccupation with the two-person zero-sum game." As weapons became ever more destructive, all-out war could not be seen as a situation of pure conflict in which opponents shared no common interests. Nash's model thus seemed more promising than von Neumann's.

Probably the single most important work Nash did at RAND involved an experiment. Designed with a team that included Milnor and published as "Some Experimental *n*-Person Games," it anticipated by several decades the now-thriving field of experimental economics. At the time the experiment was regarded as a failure, Alvin Roth has pointed out, casting doubt on the predictive power of game theory. But it later became a model because it drew attention to two aspects of interaction. First, it highlighted the importance of information possessed by participants. Second, it revealed that players' decisions were, more often than not, motivated by concerns about fairness. Despite the experiment's simplicity, it showed that watching how people actually play a game drew researchers' attention to elements of interaction—such as signaling and implied threats—that weren't part of the original model.

Nash, whose own interests were rapidly shifting away from game theory to pure mathematics, became fascinated with computers at RAND. Of the dozen or so working papers he wrote during his summers in Santa Monica, none is more visionary than one, written in his last summer at the think tank, called "Parallel Control" (see chapter 9).

Nash, however, was bent on proving himself a pure mathematician. Even before completing his thesis on game theory, he turned his attention to the trendy topic of geometric objects called manifolds. Manifolds play a role in many physical problems, including cosmology. Right off the bat, he made what he called "a nice discovery relating to manifolds and real algebraic varieties." Hoping for an appointment at Princeton or another prestigious math department, he returned to Princeton for a post-doctoral year and devoted himself to working out the details of the difficult proof.

Many breakthroughs in mathematics come from seeing unsuspected connections between objects that appear intractable and ones that mathematicians have already got their arms around. Dismissing conventional wisdom, Nash argued that manifolds were closely related to a simpler class of objects called algebraic varieties. Loosely speaking, Nash asserted that for any manifold it was possible to find an algebraic variety one of whose parts corresponded in some essential way to the original object. To do this, he showed, one has to go to higher dimensions.

Nash's theorem was initially greeted with skepticism. Experts found the notion that every manifold could be described by a system of polynomial equations implausible. "I didn't think he would get anywhere," said his Princeton adviser.

Nash completed "Real Algebraic Manifolds," his favorite paper and the only one he concedes is nearly perfect, in the fall of 1951 (see chapter 10). Its significance was instantly recognized. "Just to conceive the theorem was remarkable," said Michael Artin, a mathematician at MIT. Artin and Barry Mazur, who was a student of Nash's at MIT, later used Nash's result to resolve a basic problem in dynamics, the estimation of periodic points. Artin and Mazur proved that any smooth map from a compact manifold to itself could be approximated by a smooth map such that the number of periodic points of period p grows at most exponentially with p. The proof relied on Nash's work by translating the dynamic problem into an algebraic one of counting solutions to polynomial equations.

Nonetheless, Nash's hoped-for appointment at Princeton did not materialize. Instead, he got an offer at MIT, then still the nation's leading engineering school but not the great research university that it was to become.

In 1955, Nash unveiled a stunning result to a disbelieving audience at the University of Chicago. "I did this because of a bet," he announced.

One of his colleagues at MIT had, two years earlier, challenged him. "If you're so good, why don't you solve the embedding problem . . . ?" When Nash took up the challenge and announced that "he had solved it, modulo details," the consensus around Cambridge was that "he is getting nowhere."

The precise question that Nash was posing—"Is it possible to embed any Riemannian manifold in a Euclidian space?"—was a challenge that had frustrated the efforts of eminent mathematicians for three-quarters of a century.

By the early 1950s, interest had shifted to geometric objects in higher dimensions, partly because of the large role played by distorted-time and space relationships in Einstein's theory of relativity. Embedding means presenting a given geometric object as a subset of a space of possibly higher dimension, while preserving its essential topological properties. Take, for instance, the surface of a balloon, which is two-dimensional. You cannot put it on a blackboard, which is two-dimensional, but you can make it a subset of a space of three or more dimensions.

John Conway, the Princeton mathematician who discovered surreal numbers, calls Nash's result "one of the most important pieces of mathematical analysis in this century." Nash's theorem stated that any kind of surface that embodied a special notion of smoothness could actually be embedded in a Euclidean space. He showed, essentially, that you could fold a manifold like a handkerchief without distorting it. Nobody would have expected Nash's theorem to be true. In fact, most people who heard the result for the first time couldn't believe it. "It took enormous courage to attack these problems," said Paul Cohen, a mathematician who knew Nash at MIT.

After the publication of "The Imbedding Problem for Riemannian Manifolds" in the *Annals of Mathematics* (see chapter 11), the earlier perspective on partial differential equations was completely altered. "Many of us have the power to develop existing ideas," said Mikhail Gromov, a geometer whose work was influenced by Nash. "We follow paths prepared by others. But most of us could never produce anything comparable to what Nash produced. It's like lightening striking . . . there has been some tendency in recent decades to move from harmony to chaos. Nash said that chaos was just around the corner."

Nominally attached to the Institute for Advanced Study during a leave from MIT in the academic year 1956–57, Nash instead gravitated to the Courant Institute at New York University, "the national capital of applied mathematical analysis."

At Courant, then housed in a former hat factory off Washington Square in Greenwich Village, a group of young mathematicians was responsible for the rapid progress stimulated by World War II in the field of partial differential equations. Such equations were useful in modeling a wide variety of physical phenomena, from air passing under the wings of a jet to heat passing through metal. By the mid-1950s, mathematicians knew simple routines for solving ordinary differential equations using computers. But straightforward methods for solving most nonlinear partial differential equations—the kind potentially useful for describing large or abrupt changes—did not exist. Stanislaw Ulam complained that such systems of equations were "baffling analytically," noting that they defied "even qualitative insights by present methods."

Nash proved basic local existence, uniqueness, and continuity theorems (and also speculated about relations with statistical mechanics, singularities, and turbulence.) He used novel methods of his own invention. He had a theory that deep problems wouldn't yield to a frontal attack. Taking an ingeniously roundabout approach, he first transformed the non-linear equations into linear ones and then attacked them with non-linear means. Today rocket scientists on Wall Street use Nash-inspired methods for solving a particular class of parabolic partial differential equations that arise in finance problems.

When he returned to MIT the following fall, there were still gaps in the proof. "It was as if he was a composer and could hear the music, but he didn't know how to write it down." Nash organized a cadre of mathematicians to help him get the paper ready for publication. "It was like building the atom bomb . . . a kind of factory," said one of them later. The complete proof was published in 1958 in "Continuity of Solutions of Parabolic and Elliptic Equations" (see chapter 12).

As Nash's thirtieth birthday approached, he seemed poised to make more groundbreaking contributions. He told colleagues of "an idea of an idea" about a possible solution to the Riemann hypothesis, the deepest puzzle in all of mathematics. He set out "to revise quantum theory," along lines he had once, as a first-year graduate student, described to Einstein. Writing to Oppenheimer in 1957, Nash had said, "To me one of the best things about the Heisenberg paper is its restriction to observable quantities . . . I want to find a different and more satisfying under-picture of a non-observable reality."

Later, he blamed the onset of his terrible disease on intellectual over-reaching. No one can know what he might have accomplished had full-blown schizophrenia not set in. In the event, despite the ravages of his illness, he did go on to publish several more papers. "Le problème de Cauchy pour les équations différentielles d'une fluide générale," which appeared in 1962, is described as "basic and noteworthy" in *The Encyclopedic Dictionary of Mathematics* and inspired a good deal of subsequent work by others. He continued to tackle new subjects. Hironaka eventually wrote up one of his conjectures, dating from 1964, as "Nash Blowing Up." In 1966, he published "Analyticity of Solutions of Implicit Function Problems with Analytic Data," which pursued his ideas about partial differential equations to their natural conclusion. And in 1967 he completed a much-cited draft, "Arc Structure of Singularities," that was eventually published in a 1995 special issue of the *Duke Journal of Mathematics*.

"If you're going to develop exceptional ideas, it requires a type of thinking that is not simply practical thinking," Nash told a reporter recently. When Nash won the Nobel in 1994, he was not invited to deliver the customary hour-long Nobel lecture in Stockholm. He did, however, give a talk in Uppsala just after the Nobel ceremonies about his recent attempt to develop a mathematically correct theory of a non-expanding universe that is consistent with known physical observations. More recently, Nash has been working on game theory again. He has received a grant from the National Science Foundation to develop a new "evolutionary" solution concept for cooperative games. To get your life back is a marvelous thing, he has said. But to be able to create exciting new mathematics is now, as ever, his greatest ambition.

References

Dixit, A., and B. Nalebuff. *Thinking Strategically.* New York: W. W. Norton, 1991.

Kuhn, H. W. "Introduction," to *A Celebration of John F. Nash, Jr.,* ed. H. W. Kuhn, L. Nirenberg, and P. Sarnak, pp. i–v. In *Duke Mathematical Journal,* 81, nos. 1 and 2 (1995).

———. "Foreword." In *Classics in Game Theory,* ed. H. W. Kuhn, pp. ix–x. Princeton: Princeton University Press, 1997.

Milnor, J. "A Nobel Prize for John Nash." *The Mathematical Intelligencer* 17, no. 3 (1995): 11–17.

Myerson, R. B. "Nash Equilibrium and the History of Economic Theory." *Journal of Economic Literature* 37 (1999): 1067–82.

Nasar, S. "The Lost Years of the Nobel Laureate," *New York Times,* November 13, 1994, sec. F, pp. 1, 8.

———. *A Beautiful Mind.* New York: Simon & Schuster, 1998.

Roth, A. "Game Theory as a Tool for Market Design" (1999). Available at http://www.economics.harvard.edu/˜aroth/design.pdf.

Rubinstein, A. "John Nash: The Master of Economic Modeling." *The Scandinavian Journal of Economics* 97, no. 1 (1995): 9–13.

THE ESSENTIAL **JOHN NASH**

CHAPTER I

Press Release—The Royal Swedish Academy of Sciences

11 October 1994

The Royal Swedish Academy of Sciences has decided to award the Bank of Sweden Prize in Economic Sciences in Memory of Alfred Nobel, 1994, jointly to

Professor **John C. Harsanyi,** University of California, Berkeley, CA, USA,

Dr. **John F. Nash,** Princeton University, Princeton, NJ, USA,

Professor Dr. **Reinhard Selten,** Rheinische Friedrich-Willhelms-Universität, Bonn, Germany,

for their pioneering analysis of equilibria in the theory of non-cooperative games.

Games as the Foundation for Understanding Complex Economic Issues

Game theory emanates from studies of games such as chess or poker. Everyone knows that in these games, players have to think ahead—devise a strategy based on expected countermoves from the other player(s). Such strategic interaction also characterizes many economic situations, and game theory has therefore proved to be very useful in economic analysis.

The foundations for using game theory in economics were introduced in a monumental study by John von Neumann and Oskar Morgenstern entitled *Theory of Games and Economic Behavior* (1944). Today, 50 years later, game theory has become a dominant tool for analyzing economic issues. In particular, non-cooperative game theory (i.e., the branch of game theory that excludes binding agreements) has had great impact on economic research. The principal aspect of this theory is the concept of equilibrium, which is used to make predictions about the outcome of strategic interaction. John F. Nash, Reinhard Selten, and John C. Harsanyi are three researchers who have made eminent contributions to this type of equilibrium analysis.

John F. Nash introduced the distinction between cooperative games, in which binding agreements can be made, and non-cooperative games, where binding agreements are not feasible. Nash developed an equilibrium concept for non-cooperative games that later came to be called Nash equilibrium.

Reinhard Selten was the first to refine the Nash equilibrium concept for analyzing dynamic strategic interaction. He has also applied these refined concepts to analyses of competition with only a few sellers.

John C. Harsanyi showed how games of incomplete information can be analyzed, thereby providing a theoretical foundation for a lively field of research—the economics of information—which focuses on strategic situations where different agents do not know each other's objectives.

. . .

John Nash arrived at Princeton University in 1948 as a young doctoral student in mathematics. The results of his studies are reported in his doctoral dissertation entitled *Non-Cooperative Games* (1950). The thesis gave rise to "Equilibrium Points in *n*-Person Games" (*Proceedings of the National Academy of Sciences*, USA, 1950), and to an article entitled "Non-Cooperative Games" (*Annals of Mathematics*, 1951).

In his dissertation, Nash introduced the distinction between cooperative and non-cooperative games. His most important contribution to the theory of non-cooperative games was to formulate a universal solution concept with an arbitrary number of players and arbitrary preferences (i.e., not solely for two-person zero-sum games). This solution concept later came to be called Nash equilibrium. In a Nash equilibrium, all of the players' expectations are fulfilled and their chosen strategies are optimal. Nash proposed two interpretations of the equilibrium concept: one based on rationality and the other on statistical populations. According to the rationalistic interpretation, the players are perceived as rational and they have complete information about the structure of the game, including all of the players' preferences regarding possible outcomes, where this information is common knowledge. Since all players have complete information about each other's strategic alternatives and preferences, they can also compute each other's optimal choice of strategy for each set of expectations. If all of the players expect the same Nash equilibrium, then there are no incentives for anyone to change his strategy. Nash's second interpretation—in terms of statistical populations—is useful in so-called evolutionary games. This type of game has also been developed in biology in order to understand how the principles of natural selection operate in strategic interaction within and among species. Moreover, Nash showed that for every game with a finite number of players, there exists an equilibrium in mixed strategies.

· · ·

Through their contributions to equilibrium analysis in non-cooperative game theory, the three laureates constitute a natural combination: **Nash** provided the foundations for the analysis, while **Selten** developed it with respect to dynamics, and **Harsanyi** with respect to incomplete information.

CHAPTER 2

J O H N   F .   N A S H ,   J R .

Autobiography

My beginning as a legally recognized individual occurred on June 13, 1928 in Bluefield, West Virginia, in the Bluefield Sanitarium, a hospital that no longer exists. Of course I can't consciously remember anything from the first two or three years of my life after birth. (And, also, one suspects, psychologically, that the earliest memories have become "memories of memories" and are comparable to traditional folk tales passed on by tellers and listeners from generation to generation.) But facts are available when direct memory fails for many circumstances.

My father, for whom I was named, was an electrical engineer and had come to Bluefield to work for the electrical utility company there which was and is the Appalachian Electric Power Company. He was a veteran of WWI and had served in France as a lieutenant in the supply services and consequently had not been in actual front lines combat in the war. He was originally from Texas and had obtained his B.S. degree in electrical engineering from Texas Agricultural and Mechanical (Texas A. and M.).

My mother, originally Margaret Virginia Martin, but called Virginia, was herself also born in Bluefield. She had studied at the University of West Virginia and was a school teacher before her marriage, teaching English and sometimes Latin. But my mother's later life was considerably affected by a partial loss of hearing resulting from a scarlet fever infection that came at the time when she was a student at WVU.

Her parents had come as a couple to Bluefield from their original homes in western North Carolina. Her father, Dr. James Everett Martin, had prepared as a physician at the University of Maryland in Baltimore and came to Bluefield, which was then expanding rapidly in population, to start up his practice. But in his later years Dr. Martin became more of a real estate investor and left actual medical practice. I never saw my grandfather because he had died before I was born but I have good memories of my grandmother and of how she could play the piano at the old house which was located rather centrally in Bluefield.

A sister, Martha, was born about two and a half years later than me, on November 16, 1930.

I went to the standard schools in Bluefield but also to a kindergarten before starting in the elementary school level. And my parents provided an encyclopedia, Compton's Pictured Encyclopedia, that I learned a lot from by reading it as a child. And also there were other books available from either our house or the house of the grandparents that were of educational value.

Bluefield, a small city in a comparatively remote geographical location in the Appalachians, was not a community of scholars or of high technology. It was a center of businessmen, lawyers, etc. that owed its existence to the railroad and the rich nearby coal fields of West Virginia and western Virginia. So, from the intellectual viewpoint, it offered the sort of challenge that one had to learn from the world's knowledge rather than from the knowledge of the immediate community.

By the time I was a student in high school I was reading the classic "Men of Mathematics" by E. T. Bell and I remember succeeding in proving the classic Fermat theorem about an integer multiplied by itself p times where p is a prime.

I also did electrical and chemistry experiments at that time. At first, when asked in school to prepare an essay about my career, I prepared one about a career as an electrical engineer like my father. Later, when I actually entered Carnegie Tech. in Pittsburgh I entered as a student with the major of chemical engineering.

Regarding the circumstances of my studies at Carnegie (now Carnegie Mellon U.), I was lucky to be there on a full scholarship, called the George Westinghouse Scholarship. But after one semester as a chem. eng. student I reacted negatively to the regimentation of courses such as mechanical drawing and shifted to chemistry instead. But again, after continuing in chemistry for a while I encountered difficulties with quantitative analysis where it was not a matter of how well one could think and understand or learn facts but of how well one could handle a pipette and perform a titration in the laboratory. Also the mathematics faculty were encouraging me to shift into mathematics as my major and explaining to me that it was not almost impossible to make a good career in America as a mathematician. So I shifted again and became officially a student of mathematics. And in the end I had learned and progressed so much in mathematics that they gave me an M.S. in addition to my B.S. when I graduated.

I should mention that during my last year in the Bluefield schools my parents had arranged for me to take supplementary math. courses at Bluefield College, which was then a 2-year institution operated by Southern Baptists. I didn't get official advanced standing at Carnegie because of my extra studies but I had advanced knowledge and ability and didn't need to learn much from the first math. courses at Carnegie.

When I graduated I remember that I had been offered fellowships to enter as a graduate student at either Harvard or Princeton. But the Princeton fellowship was somewhat more generous since I had not actually won the Putnam competition and also Princeton seemed more interested in getting me to come there. Prof. A. W. Tucker wrote a letter to me encouraging me to come to Princeton and from the family point of view it seemed attractive that geographically Princeton was much nearer to Bluefield. Thus Princeton became the choice for my graduate study location.

But while I was still at Carnegie I took one elective course in "International Economics," and as a result of that exposure to economic ideas and problems arrived at the idea that led to the paper "The Bargaining

Problem" which was later published in Econometrica. And it was this idea which in turn, when I was a graduate student at Princeton, led to my interest in the game theory studies there which had been stimulated by the work of von Neumann and Morgenstern.

As a graduate student I studied mathematics fairly broadly and I was fortunate enough, besides developing the idea which led to "Non-Cooperative Games," also to make a nice discovery relating to manifolds and real algebraic varieties. So I was prepared actually for the possibility that the game theory work would not be regarded as acceptable as a thesis in the mathematics department and then that I could realize the objective of a Ph.D. thesis with the other results.

But in the event the game theory ideas, which deviated somewhat from the "line" (as if of "political party lines") of von Neumann and Morgenstern's book, were accepted as a thesis for a mathematics Ph.D. and it was later, while I was an instructor at M.I.T., that I wrote up "Real Algebraic Manifolds" and sent it in for publication.

I went to M.I.T. in the summer of 1951 as a "C.L.E. Moore Instructor." I had been an instructor at Princeton for one year after obtaining my degree in 1950. It seemed desirable more for personal and social reasons than academic ones to accept the higher-paying instructorship at M.I.T.

I was on the mathematics faculty at M.I.T. from 1951 until I resigned in the spring of 1959. During the academic year 1956–1957 I had an Alfred P. Sloan grant and chose to spend the year as a (temporary) member of the Institute for Advanced Study in Princeton.

During this period of time I managed to solve a classical unsolved problem relating to differential geometry which was also of some interest in relation to the geometric questions arising in general relativity. This was the problem to prove the isometric embeddability of abstract Riemannian manifolds in flat (or "Euclidean") spaces. But this problem, although classical, was not much talked about as an outstanding problem. It was not like, for example, the 4-color conjecture.

So as it happened, as soon as I heard in conversation at M.I.T. about the question of the embeddability being open I began to study it. The first break led to a curious result about the embeddability being realizable in surprisingly low-dimensional ambient spaces provided that one would accept that the embedding would have only limited smoothness. And later, with "heavy analysis," the problem was solved in terms of embeddings with a more proper degree of smoothness.

While I was on my "Sloan sabbatical" at the IAS in Princeton I studied another problem involving partial differential equations which I had learned of as a problem that was unsolved beyond the case of 2 dimensions. Here, although I did succeed in solving the problem, I ran into some bad luck since, without my being sufficiently informed on what other people were doing in the area, it happened that I was working in parallel with Ennio de Giorgi of Pisa, Italy. And de Giorgi was first actually to achieve the ascent of the summit (of the figuratively described problem), at least for the particularly interesting case of "elliptic equations."

It seems conceivable that if either de Giorgi or Nash had failed in the attack on this problem (of a priori estimates of Hölder continuity) then the lone climber reaching the peak would have been recognized with mathematics' Fields medal (which has traditionally been restricted to persons less than 40 years old).

Now I must arrive at the time of my change from scientific rationality of thinking into the delusional thinking characteristic of persons who are psychiatrically diagnosed as "schizophrenic" or "paranoid schizophrenic." But I will not really attempt to describe this long period of time but rather avoid embarrassment by simply omitting to give the details of truly personal type.

While I was on the academic sabbatical of 1956–1957 I also entered into marriage. Alicia had graduated as a physics major from M.I.T. where we had met and she had a job in the New York City area in 1956–1957. She had been born in El Salvador but came at an early age to the U.S. and she and her parents had long been U.S. citizens, her father being an M.D. and ultimately employed at a hospital operated by the federal government in Maryland.

The mental disturbances originated in the early months of 1959 at a time when Alicia happened to be pregnant. And as a consequence I resigned my position as a faculty member at M.I.T. and ultimately, after spending 50 days under "observation" at the McLean Hospital, travelled to Europe and attempted to gain status there as a refugee.

I later spent times of the order of five to eight months in hospitals in New Jersey, always on an involuntary basis and always attempting a legal argument for release.

And it did happen that when I had been long enough hospitalized I would finally renounce my delusional hypotheses and revert to thinking of myself as a human of more conventional circumstances and return to mathematical research. In these interludes of, as it were, enforced rationality, I did succeed in doing some respectable mathematical research. Thus there came about the research for "Le Problème de Cauchy pour les Équations Différentielles d'un Fluide Générale"; the idea that Prof. Hironaka called "the Nash blowing-up transformation"; and those of "Arc Structure of Singularities" and "Analyticity of Solutions of Implicit Function Problems with Analytic Data."

But after my return to the dream-like delusional hypotheses in the later '60s I became a person of delusionally influenced thinking but of relatively moderate behavior and thus tended to avoid hospitalization and the direct attention of psychiatrists.

Thus further time passed. Then gradually I began to intellectually reject some of the delusionally influenced lines of thinking which had been characteristic of my orientation. This began, most recognizably, with the rejection of politically-oriented thinking as essentially a hopeless waste of intellectual effort.

So at the present time I seem to be thinking rationally again in the style that is characteristic of scientists. However this is not entirely a matter of joy as if someone returned from physical disability to good physical health. One aspect of this is that rationality of thought imposes a limit on a person's concept of his relation to the cosmos. For example, a non-Zoroastrian could think of Zarathustra as simply a madman who led millions of naive followers to adopt a cult of ritual fire worship. But without his "madness" Zarathustra would necessarily have been only another of the millions or billions of human individuals who have lived and then been forgotten.

Statistically, it would seem improbable that any mathematician or scientist, at the age of 66, would be able through continued research efforts, to add much to his or her previous achievements. However I am still making the effort and it is conceivable that with the gap period of about 25 years of partially deluded thinking providing a sort of vacation my situation may be atypical. Thus I have hopes of being able to achieve something of value through my current studies or with any new ideas that come in the future.

JOHN F. NASH, JR.

Autobiography

I

"This man is a genius," read a one-line letter of recommendation introducing John Forbes Nash, Jr. to Princeton's elite mathematics department. Fourteen months after his arrival, Nash was writing the twenty-seven-page doctoral thesis that would one day revolutionize economic theory and win him a Nobel. In June 1950 the newly minted Ph.D. was about to head west to RAND, the cold war think tank.

2

As a graduate student, Nash learned mathematics not by reading or going to class but by trying out his own ideas. Physics was one of his interests. Shortly after his arrival in Princeton, he sought out Albert Einstein—standing here in the study of the famous house on Mercer Street—and regaled the saintly scientist with a theory involving "gravity, friction and radiation." Einstein listened patiently and then said, "You had better study some more physics, young man." Decades later, Nash's idea was taken up by a German physicist.

3

In the late 1940s the theory of games was very much in vogue at Princeton, where mathematicians hoped it would do for economics what Newton's calculus had done for physics. Nash soon became a regular at the game theory seminar in Princeton's mathematics department. Oskar Morgenstern, a patrician Austrian émigré economist and co-author with John von Neumann of *Theory of Games and Economic Behavior* (1944), encouraged Nash to publish a paper on bargaining.

4

John von Neumann, the Hungarian polymath (left) who worked on the atom bomb during World War II, and Robert Oppenheimer (right), the physicist who led the Manhattan Project, were among the "popes" of twentieth-century science who were in Princeton in the late 1940s. Here, they are standing in front of MANIAC, the first of the computers designed by von Neumann and his associates. Von Neumann, who felt the future of game theory was in cooperative theory, dismissed Nash's theory of non-cooperative games, calling the Nash equilibrium "just another fixed point theorem."

5

Albert W. Tucker, an authority on linear programming who nurtured pioneers of artificial intelligence like Marvin Minsky and John McCarthy, became Nash's thesis adviser. Nash called his mentor "The Machine" and initially resisted his demand to include a specific application of his equilibrium concept. Lloyd Shapley, a fellow graduate student, helped break the deadlock by supplying Nash with a nifty example involving poker.

6

John Milnor, the topologist, was arguably the most famous freshman in Princeton history and one of the few students with whom Nash liked talking mathematics. Milnor won instant fame when, as an undergraduate, he solved Borsuk's conjecture, a famous problem in knot theory, after supposedly mistaking it for a homework assignment in a class with Tucker. A major influence on Nash, Milnor, pictured here in the 1980s at the Institute for Advanced Study, was among those who helped Nash find work when he became ill.

7

David Gale, who entered Princeton as a graduate student after wartime research at the Radiation Laboratory at MIT, was the first to see the significance of the Nash equilibrium. A lifelong devotee of mathematical games, he made the first board on which the game of "Nash" was played in the Fine Hall common room. More importantly, Gale urged Nash "to plant a flag" for his non-cooperative theory by announcing his novel result in the *Proceedings of the National Academy of Sciences.* "I certainly knew right away that it was a thesis," Gale said years later. "I didn't know it was a Nobel."

8

Harold Kuhn, one of the authors of the Kuhn-Tucker theorem, was a fellow graduate student and one of the brilliant circle around Nash that included John Milnor, Lloyd Shapley, John McCarthy, and David Gale. Later, Kuhn, as Scientific Director of MATHEMATICA, a Princeton-based consulting firm, hired John Harsanyi and Reinhard Selten, Nash's fellow Nobelists, for a project that applied game theory to disarmament. Kuhn and Nash remain friends to this day. It was Kuhn who would tell Nash— the day before the final vote and official announcement—"You have won a Nobel prize."

9

Thanks more to "a nice discovery relating to manifolds and real algebraic varieties" than to his game theory thesis, Nash was made Moore Instructor at MIT and later won a faculty appointment. MIT was still more of an engineering school than one of the nation's premier research universities, but it boasted brilliant refugees, like Norbert Wiener, the father of cybernetics, and Paul Samuelson, the economist, courtesy of the anti-Semitic policies of more elite institutions like Harvard. Nash imported Princeton traditions like tea, games, and outrageous rivalry to MIT's mathematics common room.

10

Norbert Wiener, one of the most eminent American mathematicians of the pre-war generation, father of cybernetics and author of the bestselling autobiography *I Am a Mathematician*, was one of Nash's few heroes. Nash mimicked Wiener's eccentric mannerisms, including his habit of walking holding onto the walls.

II

Nash liked to drag race on Memorial Drive
in his blue 1951 Olds convertible. By the
time he joined the MIT faculty, his interest
in game theory had faded and he took up
the notoriously difficult "embedding problem" of Riemannian manifolds. "I did this
because of a bet," he told an audience after
he solved the problem.

12

The year that the Russians launched Sputnik, Nash got to know a brilliant refugee from East Germany named Jürgen Moser. Moser subsequently devised a simpler proof for Nash's embedding result—known as the Nash-Moser Implicit Function Theorem—and applied it in a general theory on celestial mechanics regarding the stability of the solar system, providing a basis for significant advances in basic space science.

13

In 1957, after Nash made a breakthrough in the field of partial differential equations that he (rightly) believed would make him a contender for the coveted Fields medal—the equivalent of a Nobel in mathematics—he was shocked to discover that an unknown Italian, Ennio de Giorgi, had beaten him to the punch. To be sure, Nash's work was independent of de Giorgi's, and the methods he employed were highly novel. Lars Garding, an expert in this area, declared, "You have to be a genius to do that." But Nash always believed de Giorgi's achievement cost him the Fields. When the two met, another mathematician said, "It was like Stanley meeting Livingstone."

14

Alicia Nash was a student in one of Nash's classes at MIT. A physics major from El Salvador in an era that worshipped dumb blondes, she thought Nash looked like Rock Hudson and believed he would one day be famous. She and Nash were married in 1957.

15

By his thirtieth birthday, Nash seemed to have it all: he was working on the Riemann hypothesis, was about to be promoted to full professor, and was singled out by *Fortune* magazine as one of the most promising mathematicians of the younger generation. Alicia was expecting a baby and Nash was planning a sabbatical in Europe. Within months, however, he was suffering from sleeplessness, paranoia, and, increasingly, bizarre delusions that he was the Prince of Peace and the Emperor of Antarctica. In May 1959 he was diagnosed with paranoid schizophrenia and resigned from MIT.

During the three decades that Nash would refer to as "the lost years," he endured repeated hospitalizations. Except for several brief interludes, he abandoned mathematics and devoted himself to numerology. Known around Princeton as "The Phantom," he survived because of the loyalty of his wife, Alicia, pictured here in Princeton with their young son, John Charles Martin Nash. The Nashes divorced in 1963, but in 1970 Alicia Nash took her ex-husband back into her home.

17

Nash's older son, John David Stier, about twenty-five years old here, grew up in Boston with his mother, Eleanor, and attended Amherst College. Father and son spent many years out of touch, but John David contacted his father and visited him in Princeton in the mid-1970s. They are once again close.

18

Nash's younger son, John Charles, pictured here at age fifteen, grew up in Princeton and became a mathematician like his father, eventually receiving a Ph.D. from Rutgers. He is also a chess master.

By the summer of 1993, rumors were flying of a possible Nobel memorial prize in Economic Sciences for game theory. In October 1994, the Swedish Academy of Sciences announced that Nash, John Harsanyi (to left in picture below) of the University of California at Berkeley and Reinhard Selten of the University of Bonn (on right) would be honored for their contributions to non-cooperative game theory. Nash was honored for his equilibrium concept. Harsanyi and Selten extended Nash's idea in crucial ways. Selten devised a theory for discriminating between plausible and implausible equilibria, while Harsanyi showed how Nash's concept could be applied in a world of less than perfect information.

20

Assar Lindbeck, chairman of the economics Nobel prize committee, called this scene of Nash bowing to the King of Sweden during the Nobel ceremony in December 1994 a vision from "a fairy tale."

21

Because of the Nobel committee's qualms about his history, Nash was not asked to give the customary Nobel lecture. But a member of the Swedish Academy of Sciences, whose own son suffered from schizophrenia, invited him to deliver a talk on cosmology at the University of Uppsala.

22

Nash's personal triumph inspired a profile in the
New York Times, a biography, and, most recently, a
Hollywood movie, *A Beautiful Mind*, directed by
Ron Howard and starring Russell Crowe, Ed
Harris, and Jennifer Connelly. After filming began
in Princeton in March 2001, Nash stopped by the
movie set. Crowe, the Australian star who had just
won an Oscar for his performance in *The
Gladiator*, served him a cup of tea and took the
opportunity to study Nash's expressive hands. Nash
is seen here facing Crowe and Howard.

23

On June 1, 2001, thirty-eight years after they were divorced, John and Alicia Nash took "a big step" in putting back together lives disrupted by mental illness. They were married for a second time by the mayor of Princeton Junction. *The New York Times* quoted Nash who called the event "a second take. Just like a movie."

24

Within a few years of receiving the Nobel, Nash
was awarded a grant from the National Science
Foundation to pursue research on cooperative
game theory and was once again giving seminars
and lectures. Here he is in June 2001, giving a talk
at the Institute for Advanced Study on his current
research.

Editor's Introduction to Chapter 3

The following excerpt from John Milnor's gracious tribute, "A Nobel Prize for John Nash," *The Mathematical Intelligencer*, no. 3 (1995), pp. 11–17, describes the discovery of the game of Hex and gives an elegant proof that the first player has a strategy that ensures a win. Nash's first version of the game, as told to me at the time, was different; its very clumsiness has always vouched for the fact that Nash discovered this beautiful game independently. He first conceived of the game in a polar form of a graph in which every vertex has valence six and is connected to six adjacent vertices. The paving of the plane by hexagons is mathematically equivalent and much more aesthetically pleasing.

[H.W.K.]

CHAPTER 3

JOHN MILNOR

The Game of Hex

Nash entered Princeton as a graduate school student in 1948, the same year that I entered as a freshman. I quickly got to know him, since we both spent a great deal of time in the common room. He was always full of mathematical ideas, not only on game theory, but in geometry and topology as well. However, my most vivid memory of this time is of the many games which were played in the common room. I was introduced to Go and Kriegspiel, and also to an ingenious topological game which we called Nash, in honor of the inventor. In fact it was later discovered that the same game had been invented a few years earlier by Piet Hein in Denmark. Hein called it Hex, and it is now commonly known by that name. An $n \times n$ Nash or Hex board consists of a rhombus which is tiled by n^2 hexagons, as illustrated in Figure 1. (The recommended size for an enjoyable game is 14×14. However, a much smaller board is shown here for illustrative purposes.) Two opposite edges are colored black, and the remaining two are colored white. The players alternately place pieces on the hexagons, and once played, a piece is never moved. The black player tries to construct a connected chain of black pieces joining the two black boundaries, while the white player tries to form a connected chain of white pieces joining the white boundaries. The game continues until one player or the other succeeds.

THEOREM. *On an n × n Hex board, the first player can always win.*

Nash's proof is marvelously nonconstructive and can be outlined as follows.

First Step. A purely topological argument shows that, in any play of the game, one player or the other must win: If the board is covered by black and white pieces, then there exists either a black chain from black to black or a white chain from white to white, but never both.

Second Step. Since this game is finite, with only two possible outcomes, and since the players move alternately with complete information, a theorem of Zermelo, rediscovered by von Neumann and Morgenstern, asserts that one of the two players must have a winning strategy.

Third Step, by symmetry. If the second player had a winning strategy, the first player could just make an initial move at random, and then follow the strategy for the second player. Since his initial play can never

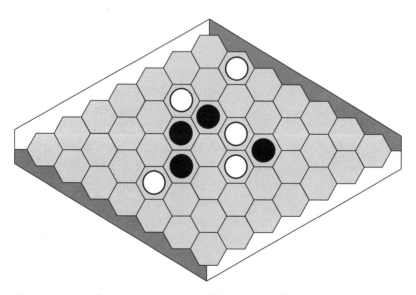

Figure 1. A typical situation in the game of Hex. Problem: Black to move and win. Alternate Problem: White to move and win.

hurt him, he must win. Thus, the hypothesis that the second player has a winning strategy leads to a contradiction. (This is a well-known argument which applies to some other symmetric games, such as Five-in-a-Row.)

Note that this proof depends strongly on the symmetry of the board. On an $n \times (n+1)$ board, the player with the shorter distance to connect can always win, even if the other player has the first move. (Compare Figure 2.)

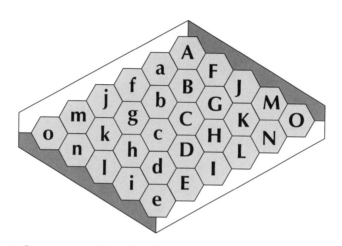

Figure 2. On an asymmetric board, as shown, white can win even if black moves first. The winning strategy can be explicitly described by "doubles": Whatever move black makes, white responds by playing in the hexagon that is marked with the corresponding symbol, where the correspondence (for example, $a \leftrightarrow A$) is a glide reflection that flips the left half of the board onto the right half.

Editor's Introduction to Chapter 4

"The Bargaining Problem" belongs with three other classic papers as a superb example of a problem treated by the axiomatic method. The other papers are Kenneth Arrow's "Impossibility Theorem," John Milnor's "Games Against Nature," and Lloyd Shapley's "A Value for n-Person Games." In each, a set of reasonable desiderata are stated as axioms. Then an important and unexpected conclusion is derived by clear and unassailable mathematical reasoning. By the precise statement of the hypotheses, the nature of the premises is opened to discussion. In Nash's paper, the most controversial axiom is the *Independence of Irrelevant Alternatives*, which continues to generate intellectual disputes to this day.

The history of this paper is still uncertain. It is my recollection that it had been sent to von Neumann during Nash's first year as a graduate student and that Nash made an appointment to remind von Neumann of its existence. In this scenario, it had been written at Carnegie Tech as a term paper in the only course in economics that Nash ever took. Nash's current memory differs from mine; in a luncheon with Roger Meyerson in 1995, he expressed the opinion that he had written the paper after his arrival at Princeton. Whatever the true history of the paper, the examples suggest that it was written by a teenager; they involve bats, balls, and penknives. What is certain is that Nash had never read the works of Cournot, Bowley, Tintner, and Fellner cited in the paper's Introduction.

[H.W.K.]

CHAPTER 4

JOHN F. NASH, JR.

The Bargaining Problem

A new treatment is presented of a classical economic problem, one which occurs in many forms, as bargaining, bilateral monopoly, etc. It may also be regarded as a nonzero-sum two-person game. In this treatment a few general assumptions are made concerning the behavior of a single individual and a group of two individuals in certain economic environments. From these, the solution (in the sense of this paper) of the classical problem may be obtained. In the terms of game theory, values are found for the game.

Introduction

A two-person bargaining situation involves two individuals who have the opportunity to collaborate for mutual benefit in more than one way. In the simpler case, which is the one considered in this paper, no action taken by one of the individuals without the consent of the other can affect the well-being of the other one.

The author wishes to acknowledge the assistance of Professors von Neumann and Morgenstern who read the original form of the paper and gave helpful advice as to the presentation.

The economic situations of monopoly versus monopsony, of state trading between two nations, and of negotiation between employer and labor union may be regarded as bargaining problems. It is the purpose of this paper to give a theoretical discussion of this problem and to obtain a definite "solution"—making, of course, certain idealizations in order to do so. A "solution" here means a determination of the amount of satisfaction each individual should expect to get from the situation, or, rather, a determination of how much it should be worth to each of these individuals to have this opportunity to bargain.

This is the classical problem of exchange and, more specifically, of bilateral monopoly as treated by Cournot, Bowley, Tintner, Fellner, and others. A different approach is suggested by von Neumann and Morgenstern in *Theory of Games and Economic Behavior*[1] which permits the identification of this typical exchange situation with a nonzero-sum two-person game.

In general terms, we idealize the bargaining problem by assuming that the two individuals are highly rational, that each can accurately compare his desires for various things, that they are equal in bargaining skill, and that each has full knowledge of the tastes and preferences of the other.

In order to give a theoretical treatment of bargaining situations, we abstract from the situation to form a mathematical model in terms of which to develop the theory.

In making our treatment of bargaining, we employ a numerical utility of the type developed in *Theory of Games* to express the preferences, or tastes, of each individual engaged in bargaining. By this means we bring into the mathematical model the desire of each individual to maximize his gain in bargaining. We shall briefly review this theory in the terminology used in this paper.

Utility Theory of the Individual

The concept of an "anticipation" is important in this theory. This concept will be explained partly by illustration. Suppose Mr. Smith knows

---

1. John von Neumann and Oskar Morgenstern, *Theory of Games and Economic Behavior*, Princeton: Princeton University Press, 1944 (Second Edition, 1947), pp. 15–31.

he will be given a new Buick tomorrow. We may say that he has a Buick anticipation. Similarly, he might have a Cadillac anticipation. If he knew that tomorrow a coin would be tossed to decide whether he would get a Buick or a Cadillac, we should say that he had a $\frac{1}{2}$ Buick, $\frac{1}{2}$ Cadillac anticipation. Thus an anticipation of an individual is a state of expectation that may involve the certainty of some contingencies and various probabilities of other contingencies. As another example, Mr. Smith might know that he will get a Buick tomorrow and think that he has half a chance of getting a Cadillac too. The $\frac{1}{2}$ Buick, $\frac{1}{2}$ Cadillac anticipation mentioned above illustrates the following important property of anticipations: if $0 \leq p \leq 1$ and A and B represent two anticipations, there is an anticipation, which we represent by $pA + (1 - p)B$, which is a probability combination of the two anticipations where there is a probability p of A and $1 - p$ of B.

By making the following assumptions we are enabled to develop the utility theory of a single individual:

1. An individual offered two possible anticipations can decide which is preferable or that they are equally desirable.
2. The ordering thus produced is transitive: if A is better than B and B is better than C then A is better than C.
3. Any probability combination of equally desirable states is just as desirable as either.
4. If A, B, and C are as in assumption (2), then there is a probability combination of A and C which is just as desirable as B. This amounts to an assumption of continuity.
5. If $0 \leq p \leq 1$ and A and B are equally desirable, then $pA + (1 - p)C$ and $pB + (1 - p)C$ are equally desirable. Also, if A and B are equally desirable, A may be substituted for B in any desirability ordering relationship satisfied by B.

These assumptions suffice to show the existence of a satisfactory utility function, assigning a real number to each anticipation of an individual. This utility function is not unique, that is, if u is such a function then so also is $au + b$, provided $a > 0$. Letting capital letters represent anticipations and small ones real numbers, such a utility function will satisfy the following properties:

(a) $u(A) > u(B)$ is equivalent to A is more desirable than B, etc.

(b) If $0 \leq p \leq 1$ then $u[pA + (1 - p)B] = pu(A) + (1 - p)u(B)$.

This is the important linearity property of a utility function.

Two-Person Theory

In *Theory of Games and Economic Behavior* a theory of n-person games is developed which includes as a special case the two-person bargaining problem. But the theory there developed makes no attempt to find a value for a given n-person game, that is, to determine what it is worth to each player to have the opportunity to engage in the game. This determination is accomplished only in the case of the two-person zero-sum game.

It is our viewpoint that these n-person games should have values; that is, there should be a set of numbers which depend continuously upon the set of quantities comprising the mathematical description of the game and which express the utility to each player of the opportunity to engage in the game.

We may define a two-person anticipation as a combination of two one-person anticipations. Thus we have two individuals, each with a certain expectation of his future environment. We may regard the one-person utility functions as applicable to the two-person anticipations, each giving the result it would give if applied to the corresponding one-person anticipation which is a component of the two-person anticipation. A probability combination of two two-person anticipations is defined by making the corresponding combinations for their components. Thus if $[A, B]$ is a two-person anticipation and $0 \leq p \leq 1$, then

$$p[A, B] + (1 - p)[C, D]$$

will be defined as

$$[pA + (1 - p)C, pB + (1 - p)D].$$

Clearly the one-person utility functions will have the same linearity property here as in the one-person case. From this point onwards when the term anticipation is used it shall mean two-person anticipation.

In a bargaining situation one anticipation is especially distinguished: this is the anticipation of no cooperation between the bargainers. It is natural, therefore, to use utility functions for the two individuals which assign the number zero to this anticipation. This still leaves each individual's utility function determined only up to multiplication by a positive real number. Henceforth any utility functions used shall be understood to be so chosen.

We may produce a graphical representation of the situation facing the two by choosing utility functions for them and plotting the utilities of all available anticipations in a plane graph.

It is necessary to introduce assumptions about the nature of the set of points thus obtained. We wish to assume that this set of points is compact and convex, in the mathematical senses. It should be convex since an anticipation which will graph into any point on a straight line segment between two points of the set can always be obtained by the appropriate probability combination of two anticipations which graph into the two points. The condition of compactness implies, for one thing, that the set of points must be bounded, that is, that they can all be inclosed in a sufficiently large square in the plane. It also implies that any continuous function of the utilities assumes a maximum value for the set at some point of the set.

We shall regard two anticipations which have the same utility for any utility function corresponding to either individual as equivalent so that the graph becomes a complete representation of the essential features of the situation. Of course, the graph is only determined up to changes of scale, since the utility functions are not completely determined.

Now since our solution should consist of *rational* expectations of gain by the two bargainers, these expectations should be realizable by an appropriate agreement between the two. Hence, there should be an available anticipation which gives each the amount of satisfaction he should expect to get. It is reasonable to assume that the two, being rational, would simply agree to that anticipation, or to an equivalent one. Hence, we may think of one point in the set of the graph as representing the solution, and also representing all anticipations that the two might agree upon as fair bargains. We shall develop the theory by giving conditions which should hold for the relationship between this solution point and the set, and from these deduce a simple condition determining the solution point. We shall consider only those cases in which there is a possibility that both individuals could gain from the situation. (This does not exclude cases where, in the end, only one individual could have benefited because the "fair bargain" might consist of an agreement to use a probability method to decide who is to gain in the end. Any probability combination of available anticipations is an available anticipation.)

Let u_1 and u_2 be utility functions for the two individuals. Let $c(S)$ represent the solution point in a set S which is compact and convex and includes the origin. We assume:

6. If α is a point in S such that there exists another point β in S with the property $u_1(\beta) > u_1(\alpha)$ and $u_2(\beta) > u_2(\alpha)$, then $\alpha \neq c(S)$.
7. If the set T contains the set S and $c(T)$ is in S, then $c(T) = c(S)$.

We say that a set S is symmetric if there exist utility operators u_1 and u_2 such that when (a, b) is contained in S, (b, a) is also contained in S; that is, such that the graph becomes symmetrical with respect to the line $u_1 = u_2$.

8. If S is symmetric and u_1 and u_2 display this, then $c(S)$ is a point of the form (a, a), that is, a point on the line $u_1 = u_2$.

The first assumption above expresses the idea that each individual wishes to maximize the utility to himself of the ultimate bargain. The third expresses equality of bargaining skill. The second is more complicated. The following interpretation may help to show the naturalness of this assumption. If two rational individuals would agree that $c(T)$ would be a fair bargain if T were the set of possible bargains, then they should be willing to make an agreement, of lesser restrictiveness, not to attempt to arrive at any bargains represented by points outside of the set S if S contained $c(T)$. If S were contained in T, this would reduce their situation to one with S as the set of possibilities. Hence $c(S)$ should equal $c(T)$.

We now show that these conditions require that the solution be the point of the set in the first quadrant where $u_1 u_2$ is maximized. We know some such point exists from the compactness. Convexity makes it unique.

Let us now choose the utility functions so that the above-mentioned point is transformed into the point $(1, 1)$. Since this involves the multiplication of the utilities by constants, $(1, 1)$ will now be the point of maximum $u_1 u_2$. For no points of the set will $u_1 + u_2 > 2$, now, since if there were a point of the set with $u_1 + u_2 > 2$ at some point on the line segment between $(1, 1)$ and that point, there would be a value of $u_1 u_2$ greater than one (see Figure 1).

We may now construct a square in the region $u_1 + u_2 \leq 2$, which

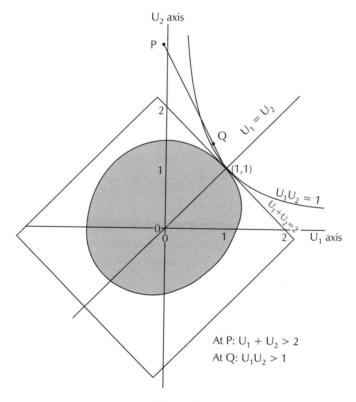

U₂ axis

P

2

Q $U_1 = U_2$

1 (1,1)

$U_1 U_2 = 1$

$U_1 + U_2 = 2$

0

1

2 U₁ axis

At P: $U_1 + U_2 > 2$
At Q: $U_1 U_2 > 1$

Figure I

is symmetrical in the line $u_1 = u_2$, which has one side on the line $u_1 + u_2 = 2$, and which completely encloses the set of alternatives. Considering the square region formed as the set of alternatives, instead of the older set, it is clear that $(1, 1)$ is the only point satisfying assumptions (6) and (8). Now using assumption (7), we may conclude that $(1, 1)$ must also be the solution point when our original (transformed) set is the set of alternatives. This establishes the assertion.

We shall now give a few examples of the application of this theory.

Examples

Let us suppose that two intelligent individuals, Bill and Jack, are in a position where they may barter goods but have no money with which to facilitate exchange. Further, let us assume for simplicity that the utility to either individual of a portion of the total number of goods involved is the sum of the utilities to him of the individual goods in that portion. We give below a table of goods possessed by each individual with the utility of each to each individual. The utility functions used for the two individuals are, of course, to be regarded as arbitrary.

Bill's goods	Utility to Bill	Utility to Jack
book	2	4
whip	2	2
ball	2	1
bat	2	2
box	4	1
Jack's goods		
pen	10	1
toy	4	1
knife	6	2
hat	2	2

The graph for this bargaining situation is included as an illustration (Figure 2). It turns out to be a convex polygon in which the point where

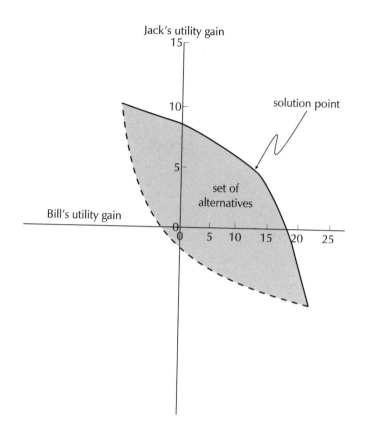

Figure 2. The solution point is on a rectangular hyperbola lying in the first quadrant and touching the set of alternatives at but one point.

the product of the utility gains is maximized is at a vertex and where there is but one corresponding anticipation. This is:

Bill gives Jack: book, whip, ball, and bat,

Jack gives Bill: pen, toy, and knife.

When the bargainers have a common medium of exchange the problem may take on an especially simple form. In many cases the money equivalent of a good will serve as a satisfactory approximate utility function. (By the money equivalent is meant the amount of money which is just as desirable as the good to the individual with

whom we are concerned.) This occurs when the utility of an amount
of money is approximately a linear function of the amount in the range
of amounts concerned in the situation. When we may use a common
medium of exchange for the utility function for each individual, the set
of points in the graph is such that the portion of it in the first quadrant
forms an isosceles right triangle. Hence the solution has each bargainer
getting the same money profit (see Figure 3).

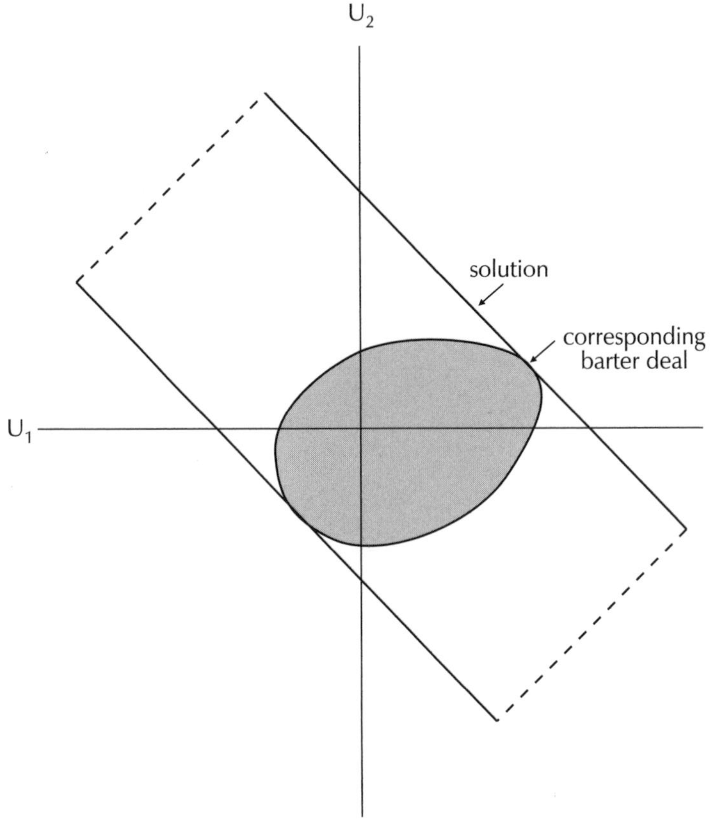

Figure 3. The inner area represents the bargains possible without the use of money.
The area between parallel lines represents the possibilities allowing the use of money.
Utility and gain measured by money are here equated for small amounts of money.
The solution must be formed using a barter-type bargain for which $u_1 + u_2$ is at a
maximum and using also an exchange of money.

Editor's Introduction to Chapters 5, 6, and 7

Two of these three contributions of Nash are titled "Non-cooperative games." Nash wanted to emphasize that he had constructed a new theory in stark contrast to the cooperative theory of von Neumann and Morgenstern. Their theory applied to two-person, non-zero-sum games and games with three or more players; and it occupies two-thirds of their book, *Theory of Games and Economic Behavior.* Nash's theory encompassed all of these cases and also two-person zero-sum games. The distinction between non-cooperative and cooperative games that Nash made is decisive to this day.

The three papers contain three different proofs of the existence of Nash equilibria and reveal an evolution in Nash's thinking. A Ph.D. thesis is understood to be solely the work of the author. In this case, however, Nash's thesis contains some joint work with Lloyd Shapley on a three-person poker game. On the other hand, the proof in the thesis of the existence of what were later called Nash equilibria is a clumsy, if totally original, application of the Brouwer fixed point theorem. At David Gale's suggestion, Kakutani's fixed point theorem was used in the note published in the *Proceedings of the National Academy of Sciences*.

It should be noted that Kakutani, who gave as an application of his theorem the proof of the minimax theorem for two-person zero-sum games, explicitly recognized the nature of this equilibrium as one in which each player played a mixed strategy that was optimal against his opponent's mixed strategy. Thus, if his argument had been given for *n*-person games, he would have shown the existence of Nash equilibria.

In my view, although it was quite natural to apply the Kakutani fixed point theorem, the most imaginative proof was provided by Nash in his paper in the *Annals of Mathematics* (previously published in a RAND memorandum). There, using a scenario in which the players adjust their strategies to give greater weight to those pure strategies that are currently best against the strategies of the remaining players, Nash gives an elegant existence proof using only Brouwer's theorem.

The Nobel selection committee apparently took the two interpretations that are contained in the thesis seriously. The rational interpretation could have been argued by Cournot, but the statistical interpretation, which is so important for biological games, is wholly original. Although the nature of non-cooperative games is explained in all three of these papers, only the thesis contains an exposition of these two interpretations. When asked at the Nobel seminar why the interpretations were not included in the *Annals* paper, Nash responded, "I don't know whether it was just pruned down in style for the *Annals of Mathematics*."

[H.W.K.]

CHAPTER 5

J O H N   F .   N A S H ,   J R .

Equilibrium Points in *n*-Person Games

Communicated by S. Lefschetz, November 16, 1949

One may define a concept of an n-person game in which each player has a finite set of pure strategies and in which a definite set of payments to the n players corresponds to each n-tuple of pure strategies, one strategy being taken for each player. For mixed strategies, which are probability distributions over the pure strategies, the pay-off functions are the expectations of the players, thus becoming polylinear forms in the probabilities with which the various players play their various pure strategies.

Any n-tuple of strategies, one for each player, may be regarded as a point in the product space obtained by multiplying the n strategy spaces of the players. One such n-tuple counters another if the strategy of each player in the countering n-tuple yields the highest obtainable expectation for its player against the $n - 1$ strategies of the other players in the countered n-tuple. A self-countering n-tuple is called an equilibrium point.

The correspondence of each n-tuple with its sets of countering

The author is indebted to Dr. David Gale for suggesting the use of Kakutani's theorem to simplify the proof and to the A. E. C. for financial support.

n-tuples gives a one-to-many mapping of the product space into itself. From the definition of countering we see that the set of countering points of a point is *convex*. By using the continuity of the pay-off functions we see that the graph of the mapping is closed. The closedness is equivalent to saying: if $P_1, P_2, \ldots$ and $Q_1, Q_2, \ldots Q_n, \ldots$ are sequences of points in the product space where $Q_n \to Q$, $P_n \to P$ and if Q_n counters P_n then Q counters P.

Since the graph is closed and since the image of each point under the mapping is convex, we infer from Kakutani's theorem[1] that the mapping has a fixed point (i.e., point contained in its image). Hence there is an equilibrium point.

In the two-person zero-sum case the "main theorem"[2] and the existence of an equilibrium point are equivalent. In this case any two equilibrium points lead to the same expectations for the players, but this need not occur in general.

1. Kakutani, S., *Duke Math. J.*, **8**, 457–459 (1941).

2. Von Neumann, J., and Morgenstern, O., *Theory of Games and Economic Behavior*, Chap. 3, Princeton University Press, Princeton, 1947.

Non-Cooperative Games

Facsimile of Ph.D. Thesis

NON-COOPERATIVE GAMES

John F. Nash, Jr.

A DISSERTATION

Presented to the Faculty of Princeton
University in Candidacy for the Degree
of Doctor of Philosophy

Recommended for Acceptance by the
Department of Mathematics

May, 1950

54

Abstract

This paper introduces the concept of a non-cooperative game and develops methods for the mathematical analysis of such games. The games considered are n-person games represented by means of pure strategies and pay-off functions defined for the combinations of pure strategies.

The distinction between cooperative and non-cooperative games is unrelated to the mathematical description by means of pure strategies and pay-off functions of a game. Rather, it depends on the possibility or impossibility of coalitions, communication, and side-payments.

The concepts of an equilibrium point, a solution, a strong solution, a sub-solution, and values are introduced by mathematical definitions. And in later sections the interpretation of these concepts in non-cooperative games is discussed.

The main mathematical result is the proof of the existence in any game of at least one equilibrium point. Other results concern the geometrical structure of the set of equilibrium points of a game with a solution, the geometry of sub-solutions, and the existence of a symmetrical equilibrium point in a symmetrical game.

As an illustration of the possibilities for application a treatment of a simple three-man poker model is included.

J O H N   F .   N A S H ,   J R .

Non-Cooperative Games

Table of Contents

Introduction

Von Neumann and Morgenstern have developed a very fruitful theory
of two-person zero-sum games in their book Theory of Games and Economic
Behavior. This book also contains a theory of n-person games of a type
which we would call cooperative. This theory is based on an analysis
of the interrelationships of the various coalitions which can be formed
by the players of the game.

Our theory, in contradistinction, is based on the absence of coalitions in that it is assumed that each participant acts independently,
without collaboration or communication with any of the others.

The notion of an equilibrium point is the basic ingredient in our
theory. This notion yields a generalization of the concept of the solution of a two-person zero-sum game. It turns out that the set of equilibrium points of a two-person zero-sum game is simply the set of all pairs
of opposing "good strategies."

In the immediately following sections we shall define equilibrium
points and prove that a finite non-cooperative game always has at least
one equilibrium point. We shall also introduce the notions of solvability
and strong solvability of a non-cooperative game and prove a theorem on
the geometrical structure of the set of equilibrium points of a solvable
game.

As an example of the application of our theory we include a solution
of a simplified three person poker game.

The motivation and interpretation of the mathematical concepts employed in the theory are reserved for discussion in a special section of
this paper.

Formal Definitions and Terminology

In this section we define the basic concepts of this paper and set up standard terminology and notation. Important definitions will be preceeded by a sub-title indicating the concept defined. The non-cooperative idea will be implicit, rather than explicit, below.

Finite Game:

For us an n-person game will be a set of n players, or positions, each with an associated finite set of pure strategies; and corresponding to each player, i, a pay-off function, P_i, which maps the set of all n-tuples of pure strategies into the real numbers. When we use the term n-tuple we shall always mean a set of n items, with each item associated with a different player.

Mixed Strategy, S_i :

A mixed strategy of player i will be a collection of non-negative numbers which have unit sum and are in one to one correspondence with his pure strategies.

We write $S_i = \sum_\alpha c_{i\alpha} \pi_{i\alpha}$ with $\sum_\alpha c_{i\alpha} = 1$ and $c_{i\alpha} \geq 0$ to represent such a mixed strategy, where the $\pi_{i\alpha}$'s are the pure strategies of player i. We regard the S_i's as points in a simplex whose vertices are the $\pi_{i\alpha}$'s. This simplex may be regarded as a convex subset of a real vector space, giving us a natural process of linear combination for the mixed strategies.

We shall use the suffixes i, j, k for players and α, β, γ to indicate various pure strategies of a player. The symbols S_i, t_i and r_i, etc. will indicate mixed strategies; $\pi_{i\alpha}$ will indicate the i th player's α th pure strategy, etc.

Pay-off function, P_i :

The pay-off function, P_i , used in the definition of a finite game above, has a unique extension to the n-tuples of mixed strategies which is linear in the mixed strategy of each player $[$n-linear$]$. This extension we shall also denote by P_i , writing $P_i(s_1, s_2, \cdots s_n)$.

We shall write $\mathfrak{L}$ or $\mathfrak{t}$ to denote an n-tuple of mixed strategies and if $\mathfrak{L} = (s_1, \cdots s_n)$ then $P_i(\mathfrak{L})$ shall mean $P_i(s_1, s_2, \cdots s_n)$ Such an n-tuple, $\mathfrak{L}$, will also be regarded as a point in a vector space, which space could be obtained by multiplying together the vector spaces containing the mixed strategies. And the set of all such n-tuples forms, of course, a convex polytope, the product of the simplices representing the mixed strategies.

For convenience we introduce the substitution notation
$$(\mathfrak{L}; t_i) \quad \text{to stand for} \quad (s_1, s_2, \cdots s_{i-1}, t_i, s_{i+1}, \cdots s_n)$$
where $\mathfrak{L} = (s_1, s_2, \cdots s_n)$. The effect of successive substitutions $((\mathfrak{L}; t_i); t_j)$ we indicate by $(\mathfrak{L}; t_i; t_j)$, etc.

Equilibrium Points:

An n-tuple $\mathfrak{L}$ is an __equilibrium point__ if and only if for every i

(1)
$$P_i(\mathfrak{L}) = \max_{\text{all } t_i's} \left[P_i(\mathfrak{L}; t_i) \right] \quad .$$

Thus an equilibrium point is an n-tuple $\mathfrak{L}$ such that each player's mixed strategy maximizes his pay-off if the strategies of the others are held fixed. Thus each player's strategy is optimal against those of the others. We shall occasionally abbreviate equilibrium point by eq. pt.

We say that a mixed strategy S_i _uses_ a pure strategy $\pi_{i\beta}$ if $S_i = \sum_\alpha c_{i\alpha} \pi_{i\alpha}$ and $c_{i\beta} > 0$. If $\mathscr{L} = (S_1, S_2, \ldots S_n)$ and S_i uses $\pi_{i\alpha}$ we also say that $\mathscr{L}$ uses $\pi_{i\alpha}$.

From the linearity of $P_i(S_1, \ldots S_n)$ in S_i,

$$(2) \qquad \max_{\text{all } r_i\text{'s}} \left[P_i(\mathscr{L} ; r_i) \right] = \max_\alpha \left[P_i(\mathscr{L} ; \pi_{i\alpha}) \right] .$$

We define $P_{i\alpha}(\mathscr{L}) = P_i(\mathscr{L} ; \pi_{i\alpha})$. Then we obtain the following trivial necessary and sufficient condition for $\mathscr{L}$ to be an equilibrium point:

$$(3) \qquad P_i(\mathscr{L}) = \max_\alpha P_{i\alpha}(\mathscr{L}) .$$

If $\mathscr{L} = (S_1, S_2, \ldots S_n)$ and $S_i = \sum_\alpha c_{i\alpha} \pi_{i\alpha}$ then $P_i(\mathscr{L}) = \sum_\alpha c_{i\alpha} P_{i\alpha}(\mathscr{L})$, consequently for (3) to hold we must have $c_{i\alpha} = 0$ whenever $P_{i\alpha}(\mathscr{L}) < \max_\beta P_{i\beta}(\mathscr{L})$, which is to say that $\mathscr{L}$ does not use $\pi_{i\alpha}$ unless it is an optimal pure strategy for player i. So we write

(4) if $\pi_{i\alpha}$ is used in $\mathscr{L}$ then $P_{i\alpha}(\mathscr{L}) = \max_\beta P_{i\beta}(\mathscr{L})$

as another necessary and sufficient condition for an equilibrium point.

Since a criterion (3) for an eq. pt. can be expressed as the equating of two continuous functions on the space of n-tuples $\mathscr{L}$ the eq. pts. obviously form a closed subset of this space. Actually, this subset is formed from a number of pieces of albegraic varieties, out out by other algebraic varieties.

I have previously published $\underline{/}$ Proc. N. A. S. 36 (1950) 48-49 $\underline{/}$ a proof of the result below based on Kakutani's generalized fixed point theorem. The proof given here uses the Brouwer theorem.

The method is to set up a sequence of continuous mappings:

$$\mathcal{L} \rightarrow \mathcal{L}'(\mathcal{L},1) \; ; \; \mathcal{L} \rightarrow \mathcal{L}'(\mathcal{L},2) \; ; \; \cdots \qquad \text{whose}$$

fixed points have an equilibrium point as limit point. A limit mapping exists, but is discontinuous, and need not have any fixed points.

THEO. 1: Every finite game has an equilibrium point.

Proof: Using our standard notation, let $\mathcal{L}$ be an n-tuple of mixed strategies, and $p_{i\alpha}(\mathcal{L})$ the pay-off to player i if he uses his pure strategy $\pi_{i\alpha}$ and the others use their respective mixed strategies in $\mathcal{L}$. For each integer λ we define the following continuous functions of $\mathcal{L}$:

$$q_i(\mathcal{L}) = \max_{\alpha} p_{i\alpha}(\mathcal{L}) \; ,$$

$$\phi_{i\alpha}(\mathcal{L},\lambda) = p_{i\alpha}(\mathcal{L}) - q_i(\mathcal{L}) + 1/\lambda \; , \text{ and}$$

$$\phi_{i\alpha}^{+}(\mathcal{L},\lambda) = \max\left[0, \phi_{i\alpha}(\mathcal{L},\lambda)\right] \; .$$

Now $\sum_{\alpha} \phi_{i\alpha}^{+}(\mathcal{L},\lambda) \geq \max_{\alpha} \phi_{i\alpha}^{+}(\mathcal{L},\lambda) = 1/\lambda > 0$ so that

$$c_{i\alpha}(\mathcal{L},\lambda) \doteq \frac{\phi_{i\alpha}^{+}(\mathcal{L},\lambda)}{\sum_{\beta} \phi_{i\beta}^{+}(\mathcal{L},\lambda)} \qquad \text{is continuous.}$$

Define $s_i'(\mathcal{L},\lambda) = \sum_{\alpha} \pi_{i\alpha} \, c_{i\alpha}(\mathcal{L},\lambda)$ and $\mathcal{L}'(\mathcal{L},\lambda) = (s_1', s_2', \cdots s_n')$. Since all the operations have preserved continuity, the mapping $\mathcal{L} \rightarrow \mathcal{L}'(\mathcal{L},\lambda)$ is con-

tinuous; and since the space of n-tuples, $\mathcal{L}$, is a cell, there must
be a fixed point for each λ . Hence there will be a subsequence $\mathcal{L}_\mu$,
converging to $\mathcal{L}^*$, where $\mathcal{L}_\mu$ is fixed under the mapping $\mathcal{L} \to \mathcal{L}'(\mathcal{L}, \lambda_{(\mu)})$.

Now suppose $\mathcal{L}^*$ were not an equilibrium point. Then if .
$$\mathcal{L}^* = (s_1^*, \ldots s_n^*)$$ some component s_i^* must be non-
optimal against the others, which means s_i^* uses some pure strategy
$\pi_{i\alpha}$ which is non-optimal. $\underline{[}$ see $(4), pg.4\underline{]}$ This means that

$$P_{i\alpha}(\mathcal{L}^*) < q_i(\mathcal{L}^*)$$ which justifies

writing $$P_{i\alpha}(\mathcal{L}^*) - q_i(\mathcal{L}^*) < -\epsilon$$.

From continuity, if μ is large enough,

$$\left|\left[P_{i\alpha}(\mathcal{L}_\mu) - q_i(\mathcal{L}_\mu)\right] - \left[P_{i\alpha}(\mathcal{L}^*) - q_i(\mathcal{L}^*)\right]\right| < \epsilon/2 \text{ and } 1/\lambda_{(\mu)} < \epsilon/2.$$

Adding, $$P_{i\alpha}(\mathcal{L}_\mu) - q_i(\mathcal{L}_\mu) + 1/\lambda_{(\mu)} < 0$$ which is
simply $$\phi_{i\alpha}(\mathcal{L}_\mu, \lambda_{(\mu)}) < 0 \text{, whence } \phi_{i\alpha}^+(\mathcal{L}_\mu, \lambda_{(\mu)}) = 0, \text{whence}$$

$$C_{i\alpha}(\mathcal{L}_\mu, \lambda_{(\mu)}) = 0$$. From this last equation we know that
$\pi_{i\alpha}$ is not used in $\mathcal{L}_\mu$ since

$$\mathcal{L}_\mu = \sum_\alpha \pi_{i\alpha} C_{i\alpha}(\mathcal{L}_\mu, \lambda_{(\mu)})$$, because $\mathcal{L}_\mu$ is a

fixed point.

And since $\mathcal{L}_\mu \to \mathcal{L}^*$, $\pi_{i\alpha}$ is not used in $\mathcal{L}^*$,
which contradicts our assumption.

Hence $\mathcal{L}^*$ is indeed an equilibrium point.

62

Symmetries of Games

An _automorphism_, or _symmetry_, of a game will be a permutation of its pure strategies which satisfies certain conditions, given below.

If two strategies belong to a single player they must go into two strategies belonging to a single player. Thus if ϕ is the permutation of the pure strategies it induces a permutation ψ of the players.

Each n-tuple of pure strategies is therefore permuted into another n-tuple of pure strategies. We may call χ the induced permutation of these n-tuples. Let ξ denote an n-tuple of pure strategies and

$$P_i(\xi)$$ the pay-off to player i when the n-tuple ξ is employed. We require that if

$$j = i^{\psi} \qquad \text{then} \qquad P_j(\xi^{\chi}) = P_i(\xi)$$

which completes the definition of a symmetry.

The permutation ϕ has a unique linear extension to the mixed strategies. If

$$S_i = \sum_{\alpha} c_{i\alpha} \pi_{i\alpha} \quad \text{we define} \quad (S_i)^{\phi} = \sum_{\alpha} c_{i\alpha}(\pi_{i\alpha})^{\phi}.$$

The extension of ϕ to the mixed strategies clearly generates an extension of χ to the n-tuples of mixed strategies. We shall also denote this by χ.

We define a _symmetric n-tuple_ $\mathscr{s}$ of a game by

$$\mathscr{s}^{\chi} = \mathscr{s} \qquad \text{for all} \quad \chi\text{'s}$$

it being understood that χ means a permutation derived from a symmetry ϕ.

THEO. 4: Any finite game has a symmetric equilibrium point.

Proof: First we note that

$$S_{i0} = \frac{\sum_{\alpha} \pi_{j\alpha}}{\sum 1}$$ has the property $(S_{i0})^\phi = S_{j0}$ where $j = i^\psi$, so that the n-tuple $\mathcal{A}_0 = (S_{10}, S_{20}, \ldots S_{n0})$ is fixed under any χ ; hence any game has at least one symmetric n-tuple.

If $\mathcal{A} = (S_1, \ldots S_n)$ and $\mathcal{A} = (t_1, \ldots t_n)$ are symmetric then $\frac{\mathcal{A}+\mathcal{A}}{2} = \left(\frac{S_1+t_1}{2}, \ldots \frac{S_n+t_n}{2}\right)$ is so too because $\mathcal{A}^\chi = \mathcal{A} \Leftrightarrow S_j = (S_i)^\phi$ where $j = i^\psi$, hence $\frac{S_j+t_j}{2} = \frac{(S_i)^\phi + (t_i)^\phi}{2} = \left(\frac{S_i+t_i}{2}\right)^\phi$, hence $\left(\frac{\mathcal{A}+\mathcal{A}}{2}\right)^\chi = \frac{\mathcal{A}+\mathcal{A}}{2}$.

This shows that the set of symmetric n-tuples is a convex subset of the space of n-tuples since it is obviously closed.

Now observe that for each λ the mapping $\mathcal{A} \to \mathcal{A}'(\mathcal{A}, \lambda)$ used in the proof of existence theorem was intrinsically defined. Therefore, if $\mathcal{A}_2 = \mathcal{A}'(\mathcal{A}_1, \lambda)$ and χ ^(a permutation derived from) is an automorphism of the game we will have $\mathcal{A}_2{}^\chi = \mathcal{A}'(\mathcal{A}^\chi, \lambda)$. If $\mathcal{A}_1$ is symmetric $\mathcal{A}_1{}^\chi = \mathcal{A}_1$ and therefore $\mathcal{A}_2{}^\chi = \mathcal{A}'(\mathcal{A}_1, \lambda) = \mathcal{A}_2$. Consequently this mapping maps the set of symmetric n-tuples into itself.

Since this set is a cell there must be a symmetric fixed point $\mathcal{A}_\lambda$. And, as in the proof of the existence theorem we could obtain a limit point $\mathcal{A}^*$ which would have to be symmetric.

Solutions

We define here solutions, strong solutions, and sub-solutions. A non-cooperative game does not always have a solution, but when it does the solution is unique. Strong solutions are solutions with special properties. Sub-solutions always exist and have many of the properties of solutions, but lack uniqueness.

S_i will denote a set of mixed strategies of player i and $\mathcal{L}$ a set of n-tuples of mixed strategies.

Solvability:

A game is solvable if its set, $\mathcal{L}$, of equilibrium points satisfies the condition

(1) $$(t; r_i) \in \mathcal{L} \quad \text{and} \quad \mathcal{2} \in \mathcal{L} \Rightarrow (\mathcal{2}; r_i) \in \mathcal{L} \quad \text{for all } i\text{'s}.$$

This is called the __interchangeability__ condition. The __solution__ of a solvable game is its set, $\mathcal{L}$, of equilibrium points.

Strong Solvability:

A game is __strongly solvable__ if it has a solution, $\mathcal{L}$, such that for all i's

$$\mathcal{2} \in \mathcal{L} \quad \text{and} \quad p_i(\mathcal{2}; r_i) = p_i(\mathcal{2}) \Rightarrow (\mathcal{2}; r_i) \in \mathcal{L}$$

and then $\mathcal{L}$ is called a __strong solution__.

Equilibrium Strategies:

In a solvable game let S_i be the set of all mixed strategies S_i

such that for some t the n-tuple $(t; s_i)$ is an equilibrium point. $\angle$ s_i is the i th component of some equilibrium point. $\angle$ We call S_i the set of equilibrium strategies of player i .

Sub-solutions:

If $\mathcal{L}$ is a subset of the set of equilibrium points of a game and satisfies condition (1); and if $\mathcal{L}$ is maximal relative to this property then we call $\mathcal{L}$ a sub-solution.

For any sub-solution $\mathcal{L}$ we define the i th factor set, S_i , as the set of all s_i's such that $\mathcal{L}$ contains $(t; s_i)$ for some t .

Note that a sub-solution, when unique, is a solution; and its factor sets are the sets of equilibrium strategies.

THEO. 2: A sub-solution, $\mathcal{L}$, is the set of all n-tuples $(s_1, s_2, \cdots s_n)$ $(s_1, s_2, \cdots s_n)$ such that each $s_i \in S_i$ where S_i is the i th factor set of $\mathcal{L}$. Geometrically, $\mathcal{L}$ is the product of its factor sets.

Proof: Consider such an n-tuple $(s_1, \cdots s_n)$. By definition $\exists \ t_1, t_2, \cdots, t_n$ such that for each i $(t_i; s_i) \in \mathcal{L}$. Using the condition (1) n-1 times we obtain successively $(t_1; s_1; s_2) \in \mathcal{L}$, $\cdots$, $(t_1; s_1; s_2; s_3; \cdots; s_n) \in \mathcal{L}$ and the last is simply $(s_1, s_2, \cdots s_n) \in \mathcal{L}$, which we needed to show.

THEO. 3: The factor sets $S_1, S_2, \cdots S_n$ of a sub-solution are closed and convex as subsets of the mixed strategy spaces.

Proof: It suffices to show two things: (a) if s_i and $s_i' \in S_i$

then $S_i^* = (S_i + S_i')/2 \in S_i$; (b) if $S_i^\#$ is
a limit point of S_i then $S_i^\# \in S_i$.

Let $t \in \mathcal{L}$. Then we have

$$P_j(t; S_i) \geq P_j(t; S_i; t_j) \quad \text{and} \quad P_j(t; S_i) \geq P_j(t; S_i'; t_j)$$

for any t_j , by using the criterion of (1) , $pg. 3$ for an eq.
pt. Adding these inequalities, using the linearity of $P_j(s_1, \ldots s_n)$ in
S_i, and dividing by 2, we get $P_j(t; S_i^*) \geq P_j(t; S_i^*; t_j)$
since $S_i^* = (S_i + S_i')/2$. From this we know that
$(t; S_i^*)$ is an eq. pt. for any $t \in \mathcal{L}$. If the set of
all such eq. pts. $(t; S_i^*)$ is added to $\mathcal{L}$ the augmented set
clearly satisfies condition (1), and since $\mathcal{L}$ was to be maximal it
follows that $S_i^* \in S_i$.

To attack (b) note that the n-tuple $(t; S_i^\#)$, where $t \in \mathcal{L}$
will be a limit point of the set of n-tuples of the form $(t; S_i)$
where $S_i \in S_i$, since $S_i^\#$ is a limit point of S_i . But
this set is a set of eq. pts. and hence any point in its closure is an
eq. pt., since the set of all eq. pts. is closed [see pg. 4] . Therefore $(t; S_i^\#)$ is an eq. pt. and hence $S_i^\# \in S_i$ from
the same argument as for S_i^* .

Values:

Let $\mathcal{L}$ be the set of equilibrium points of a game. We define

$$v_i^+ = \max_{\mathcal{L} \in \mathcal{L}} [P_i(\mathcal{L})] \quad , \quad v_i^- = \min_{\mathcal{L} \in \mathcal{L}} [P_i(\mathcal{L})] \quad .$$

If $v_i^+ = v_i^-$ we write $v_i = v_i^+ = v_i^-$. v_i^+ is
the upper value to player i of the game; v_i^- the lower value; and

V_i the value, if it exists.

Values will obviously have to exist if there is but one equilibrium point.

One can define associated values for a sub-solution by restricting $\mathcal{L}$ to the eq. pts. in the sub-solution and then using the same defining equations as above.

A two-person zero-sum game is always solvable in the sense defined above. The sets of equilibrium strategies S_1 and S_2 are simply the sets of "good" strategies. Such a game is not generally strongly solvable; strong solutions exist only when there is a "saddle point" in pure strategies.

Simple Examples

These are intended to illustrate the concepts defined in the paper and display special phenomena which occur in these games.

The first player has the roman letter strategies and the pay-off to the left, etc.

Ex. 1
$$
\begin{array}{ccc}
5 & a\,\alpha & -3 \\
-4 & a\,\beta & 4 \\
-5 & b\,\alpha & 5 \\
5 & b\,\beta & -4
\end{array}
$$

Weak Solution: $\left(\frac{9}{16}a + \frac{7}{16}b, \frac{7}{17}\alpha + \frac{10}{17}\beta\right)$

$v_1 = \frac{-5}{17}$, $v_2 = +\frac{1}{2}$

Ex. 2
$$
\begin{array}{ccc}
1 & a\,\alpha & 1 \\
-10 & a\,\beta & 10 \\
10 & b\,\alpha & -10 \\
-1 & b\,\beta & -1
\end{array}
$$

Strong Solution: (b, β)

$v_1 = v_2 = -1$

Ex. 3
$$
\begin{array}{ccc}
1 & a\,\alpha & 1 \\
-10 & a\,\beta & -10 \\
-10 & b\,\alpha & -10 \\
1 & b\,\beta & 1
\end{array}
$$

Unsolvable; equilibrium points (a,α), (b,β), and $(a/2 + b/2, \alpha/2 + \beta/2)$. The strategies in the last case have maxi-min and mini-max properties.

Ex. 4
$$
\begin{array}{ccc}
1 & a\,\alpha & 1 \\
0 & a\,\beta & 1 \\
1 & b\,\alpha & 0 \\
0 & b\,\beta & 0
\end{array}
$$

Strong Solution: all pairs of mixed strategies.

$v_1^{+} = v_2^{+} = 1$, $v_1^{-} = v_2^{-} = 0$

Ex. 5
$$
\begin{array}{ccc}
1 & a\,\alpha & 2 \\
-1 & a\,\beta & -4 \\
-4 & b\,\alpha & -1 \\
2 & b\,\beta & 1
\end{array}
$$

Unsolvable; eq. pts. (a,α), (b,β) and $(1/4\,a + 3/4\,b, 3/8\,\alpha + 5/8\,\beta)$. However, empirical tests show a tendency toward (a,α).

Ex. 6
$$
\begin{array}{ccc}
1 & a\,\alpha & 1 \\
0 & a\,\beta & 0 \\
0 & b\,\alpha & 0 \\
0 & b\,\beta & 0
\end{array}
$$

Eq. pts.: (a,α) and (b,β) , with (b,β) an example of instability.

Geometrical Form of Solutions

In the two-person zero-sum case it has been shown that the set of "good" strategies of a player is a convex polyhedral subset of his strategy space. We shall obtain the same result for a player's set of equilibrium strategies in any solvable game.

THEO. 5: The sets $S_1, S_2, \ldots S_n$ of equilibrium strategies in a solvable game are polyhedral convex subsets of the respective mixed strategy spaces.

Proof: An n-tuple $\measuredangle$ will be an equilibrium point if and only if for every i

(1) $$P_i(\measuredangle) = \max_\alpha \; P_{i\alpha}(\measuredangle)$$

which is condition (3) on page 4 . An equivalent condition is for every i and α

(2) $$P_i(\measuredangle) - P_{i\alpha}(\measuredangle) \geq 0 \;.$$

Let us now consider the form of the set S_j of equilibrium strategies, S_j , of player j . Let $\mathcal{t}$ be any equilibrium point, then $(\mathcal{t} \, ; S_j)$ will be an equilibrium point if and only if $s_j \in S_j$, from Theo. 2. We now apply conditions (2) to $(\mathcal{t} \, ; S_j)$, obtaining

(3) $s_j \in S_j \; \Longleftrightarrow$ for all $i, \alpha \quad P_i(\mathcal{t} \, ; S_j) - P_{i\alpha}(\mathcal{t} \, ; S_j) \geq 0 \;.$

Since P_i is n-linear and $\mathcal{t}$ is constant these are a set of linear inequalities of the form $F_{i\alpha}(S_j) \geq 0$. Each such inequality is either satisfied for all S_j or for those lying on and to one side of some hyperplane passing through the strategy simplex. Therefore, the

complete set $\sqrt{}$which is finite$\sqrt{}$ of conditions will all be satisfied
simultaneously on some convex polyhedral subset of player $j's$ strategy simplex. $\sqrt{}$Intersection of half-spaces.$\sqrt{}$

As a corollary we may conclude that S_k is the convex closure
of a finite set of mixed strategies $\sqrt{}$vertices$\sqrt{}$.

Dominance and Contradiction Methods

We say that S_i' dominates S_i if $p_i(t; S_i') > p_i(t; S_i)$ for every t.

This amounts to saying that S_i' gives player i a higher payoff than S_i no matter what the strategies of the other players are. To see whether a strategy S_i' dominates S_i it suffices to consider only pure strategies for the other players because of the n-linearity of p_i.

It is obvious from the definitions that no equilibrium point can involve a dominated strategy S_i.

The domination of one mixed strategy by another will always entail other dominations. For suppose S_i' dominates S_i and t_i uses all of the pure strategies which have a higher coefficient in S_i than in S_i'. Then for a small enough $\rho > 0$

$$t_i' = t_i + \rho(S_i' - S_i)$$

is a mixed strategy; and t_i' dominates t_i by linearity.

One can prove a few properties of the set of undominated strategies. It is simply connected and is formed by the union of some collection of faces of the strategy simplex.

The information obtained by discovering dominances for one player may be of relevance to the others, insofar as the elimination of classes of mixed strategies as possible components of an equilibrium point is concerned. For the t's whose components are all undominated are all that need be considered and this eliminating some of the strategies of one player may make possible the elimination of a new class of strategies for another player.

Another procedure which may be used in locating equilibrium points is the contradiction-type analysis. Here one assumes that an equilibrium point exists having component strategies lying within certain regions of the strategy spaces and proceeds to deduce further conditions which must be satisfied if the hypothesis is true. This sort of reasoning may be carried through several stages to eventually obtain a contradiction indicating that there is no equilibrium point satisfying the initial hypothesis.

A Three-Man Poker Game

As an example of the application of our theory to a more or less realistic case we include the simplified poker game given below. The rules are as follows:

(1) The deck is large, with equally many _high_ and _low_ cards, and a hand consists of one card.

(2) Two chips are used to ante, open, or call.

(3) The players play in rotation and the game ends after all have passed or after one player has opened and the others have had a chance to call.

(4) If no one bets the antes are retrieved.

(5) Otherwise the pot is divided equally among the highest hands which have bet.

We find it more satisfactory to treat the game in terms of quantities we call "behavior parameters" than in the normal form of "Theory of Games and Economic Behavior." In the normal form representation two mixed strategies of a player may be equivalent in the sense that each makes the individual choose each available course of action in each particular situation requiring action on his part with the same frequency. That is, they represent the same behavior pattern on the part of the individual.

Behavior parameters give the probabilities of taking each of the various possible actions in each of the various possible situations which may arise. Thus they describe behavior patterns.

In terms of behavior parameters the strategies of the players may be represented as follows, assuming that since there is no point in passing with a _high_ card at one's last opportunity to bet that this will not be done. The greek letters are the probabilities of the various acts.

	First Moves	Second Moves
I	α Open on <u>high</u> β Open on <u>low</u>	κ Call III on <u>low</u> λ Call II on <u>low</u> μ Call II and III on <u>low</u>
II	γ Call I on <u>low</u> δ Open on <u>high</u> ϵ Open on <u>low</u>	ν Call III on <u>low</u> ξ Call III and I on <u>low</u>
III	ς Call I and II on <u>low</u> η Open on <u>low</u> θ Call I on <u>low</u> ι Call II on <u>low</u>	Player III never gets a second move.

We locate all possible equilibrium points by first showing that most of the greek parameters must vanish. By dominance mainly with a little contradiction-type analysis β is eliminated and with it go γ, ς, and θ by dominance. Then contradictions eliminate $\mu, \xi, \iota, \lambda, \kappa$, and ν in that order. This leaves us with α, δ, ϵ and η. Contradiction analysis shows that none of these can be zero or one and thus we obtain a system of simultaneous algebraic equations. The equations happen to have but one solution with the variables in the range $(0,1)$. We get

$$\alpha = \frac{21 - \sqrt{321}}{10} \quad , \quad \eta = \frac{5\alpha + 1}{4} \quad , \quad \delta = \frac{5 - 2\alpha}{5 + \alpha} \quad , \text{ and}$$

$$\epsilon = \frac{4\alpha - 1}{\alpha + 5} . \text{ These yield } \alpha = .308, \eta = .635, \delta = .826, \text{ and}$$

$$\epsilon = .044 .$$

Since there is only one equilibrium point the game has values; these are

$$v_1 = -.147 = \frac{-(1+17\alpha)}{8(5+\alpha)}, \quad v_2 = -.096 = -\frac{1-2\alpha}{4}, \text{ and}$$

$$v_3 = .243 = \frac{79}{40}\left(\frac{1-\alpha}{5+\alpha}\right).$$

Investigation of the coalition powers yields the following "good strategies" and values for the various coalitions. Parameters not mentioned are zero.

I and II versus	III
$\alpha = 3/4$	$\iota = 1/4$, $0 \leq \eta \leq 2/3$
$\delta = \epsilon = 1$	value to III: $.03125 = 1/32$

II and III versus	I
$\delta = 1, \epsilon = 0$	$\alpha = 2/3$
$\eta = 2/3$	value to I: $-.1667 = -1/6$

I and III versus II

	high	low
bet	pass	
pass	pass	

$\cdots \eta = 0 \quad 3/11$

$\cdots \eta = 13/16 \quad 8/11$

$\delta = 7/11$, $\epsilon = 3/11$

value to II: $-.1136 = -5/44$

The coalition members have the power to agree upon a pattern of play before the game is played. This advantage becomes significant only in the case of coalition I III where III may open after two passes when I had planned to pass on both high and low but will not open if

I had planned to bet if he got high. The values given are, of course, what the single player assures himself with his "safe" strategy.

A more detailed treatment of this game is being prepared for publication elsewhere. This will consider different relative sizes of ante and bet.

Motivation and Interpretation

In this section we shall try to explain the significance of the concepts introduced in this paper. That is, we shall try to show how equilibrium points and solutions can be connected with observable phenomena.

The basic requirements for a non-cooperative game is that there should be no pre-play communication among the players $\underline{\int}$ unless it has no bearing on the game $\underline{\ }$. Thus, by implication, there are no coalitions and no side-payments. Because there is no extra-game utility $\underline{\int}$ pay-off $\underline{\ }$ transfer, the pay-offs of different players are effectively incomparable; if we transform the pay-off functions linearly: $P_i' = a_i p_i + b_i$, where $a_i > 0$ the game will be essentially the same. Note that equilibrium points are preserved under such transformations.

We shall now take up the "mass-action" interpretation of equilibrium points. In this interpretation solutions have no great significance. It is unnecessary to assume that the participants have full knowledge of the total structure of the game, or the ability and inclination to go through any complex reasoning processes. But the participants are supposed to accumulate empirical information on the relative advantages of the various pure strategies at their disposal.

To be more detailed, we assume that there is a population $\underline{\int}$ in the sense of statistics $\underline{\ }$ of participants for each position of the game. Let us also assume that the "average playing" of the game involves n participants selected at random from the n populations, and that there is a stable average frequency with which each pure strategy is employed by the "average member" of the appropriate population.

Since there is to be no collaboration between individuals playing in different positions of the game, the probability that a particular n-tuple of pure strategies will be employed in a playing of the game should be the product of the probabilities indicating the chance of each of the n pure strategies to be employed in a random playing.

Let the probability that $\pi_{i\alpha}$ will be employed in a random playing of the game be $c_{i\alpha}$, and let $S_i = \sum_\alpha c_{i\alpha} \pi_{i\alpha}$, $\mathcal{A} = (S_1, S_2, \cdots S_n)$. Then the expected pay-off to an individual playing in the ith position of the game and employing the pure strategy $\pi_{i\alpha}$ is $p_i(\mathcal{A}; \pi_{i\alpha}) = p_{i\alpha}(\mathcal{A})$.

Now let us consider what effects the experience of the participants will produce. To assume, as we did, that they accumulated empirical evidence on the pure strategies at their disposal is to assume that those playing in position i learn the numbers $p_{i\alpha}(\mathcal{A})$. But if they know these they will employ only optimal pure strategies, i.e., those pure strategies $\pi_{i\alpha}$ such that

$$p_{i\alpha}(\mathcal{A}) = \overset{max}{\underset{\beta}{}} p_{i\beta}(\mathcal{A})$$

Consequently since S_i expresses their behavior S_i attaches positive coefficients only to optimal pure strategies, so that

$$\pi_{i\alpha} \text{ is used in } S_i \implies p_{i\alpha}(\mathcal{A}) = \overset{max}{\underset{\beta}{}} p_{i\beta}(\mathcal{A})$$

But this is simply a condition for $\mathcal{A}$ to be an equilibrium point. [see (4), pg. 5]

Thus the assumptions we made in this "mass-action" interpretation lead to the conclusion that the mixed strategies representing the average behavior in each of the populations form an equilibrium point.

The populations need not be large if the assumptions still make it

. . . . . . .ill hold. There are situations in economics or international politics in which, effectively, a group of interests are involved in a non-cooperative game without being aware of it; the non-awareness helping to make the situation truly non-cooperative.

Actually, of course, we can only expect some sort of approximate euqilibrium, since the information, its utilization, and the stability of the average frequencies will be imperfect.

We now sketch another interpretation, one in which solutions play a major role, and which is applicable to a game played but once.

We proceed by investigating the question: what would be a "rational" prediction of the behavior to be expected of rational playing the game in question? By using the principles that a rational prediction should be unique, that the players should be able to deduce and make use of it, and that such knowledge on the part of each player of what to expect the others to do should not lead him to act out of conformity with the prediction, one is led to the concept of a solution defined before.

If $S_1, S_2, \ldots S_n$ were the sets of equilibrium strategies of a solvable game, the "rational" prediction should be: "The average behavior of rational men playing in position i would define a mixed strategy S_i in S_i if an experiment were carried out."

In this interpretation we need to assume the players know the full structure of the game in order to be able to deduce the prediction for themselves. It is quite strongly a rationalistic and idealizing interpretation.

In an unsolvable game it sometimes happens that good heuristic reasons can be found for narrowing down the set of equilibrium points to those in a single sub-solution, which then plays the role of a solution.

In general a sub-solution may be looked at as a set of mutually compatible equilibrium points, forming a coherent whole. The sub-solutions appear to give a natural subdivision of the set of equilibrium points of a game.

Applications

The study of n-person games for which the accepted ethics of fair play imply non-cooperative playing is, of course, an obvious direction in which to apply this theory. And poker is the most obvious target. The analysis of a more realistic poker game than our very simple model should be quite an interesting affair.

The complexity of the mathematical work needed for a complete investigation increases rather rapidly, however, with increasing complexity of the game; so that it seems that analysis of a game much more complex than the example given here would only be feasible using approximate computational methods.

A less obvious type of application is to the study of cooperative games. By a cooperative game we mean a situation involving a set of players, pure strategies, and pay-offs as usual; but with the assumption that the players can and will collaborate as they do in the von Neumann and Morgenstern theory. This means the players may communicate and form coalitions which will be enforced by an umpire. It is unnecessarily restrictive, however, to assume any transferability, or even comparability of the pay-offs $/$which should be in utility units$/$ to different players. Any desired transferability can be put into the game itself instead of assuming it possible in the extra-game collaboration.

The writer has developed a "dynamical" approach to the study of cooperative games based upon reduction to non-cooperative form. One proceeds by constructing a model of the pre-play negotiation so that the steps of negotiation become moves in a larger non-cooperative game $/$which will have an infinity of pure strategies$/$ describing the total situation. This larger game is then treated in terms of the theory of this paper [extended to infinite games] and if values are obtained they are taken as the values of the cooperative game. Thus the problem analyzing a cooperative game becomes the problem of obtaining a suitable, and convincing, non-cooperative model for the negotiation.

The writer has, by such a treatment, obtained values for all finite two person cooperative games, and some special n-person games.

Bibliography

(1) von Neumann, Morgenstern, "Theory of Games and Economic Behavior",
     Princeton University Press, 1944.

(2) J. F. Nash, Jr., "Equilibrium Points in N-Person Games", Proc. N.
     A. S. 36 (1950) 48-49.

Acknowledgements

Drs. Tucker, Gale, and Kuhn gave valuable criticism and suggestions
for improving the exposition of the material in this paper. David Gale
suggested the investigation of symmetric games. The solution of the Poker
model was a joint project undertaken by Lloyd S. Shapley and the author.
Finally, the author was sustained financially by the Atomic Energy Commission in the period 1949-50 during which this work was done.

CHAPTER 7

J O H N   F .   N A S H ,   J R .

Non-Cooperative Games

(Received October 11, 1950)

Introduction

Von Neumann and Morgenstern have developed a very fruitful theory of two-person zero-sum games in their book *Theory of Games and Economic Behavior.* This book also contains a theory of *n*-person games of a type which we would call cooperative. This theory is based on an analysis of the interrelationships of the various coalitions which can be formed by the players of the game.

Our theory, in contradistinction, is based on the *absence* of coalitions in that it is assumed that each participant acts independently, without collaboration of communication with any of the others.

The notion of an *equilibrium point* is the basic ingredient of our theory. This notion yields a generalization of the concept of the solution of a two-person zero-sum game. It turns out that the set of equilibrium points of a two-person zero-sum game is simply the set of all pairs of opposing "good strategies."

In the immediately following sections, we shall define equilibrium points and prove that a finite non-cooperative game always has at least one equilibrium point. We shall also introduce the notions of solvability and strong solvability of a non-cooperative game and prove a theorem on the geometrical structure of the set of equilibrium points of a solvable game.

As an example of the application of our theory, we include a solution of a simplified three-person poker game.

Formal Definitions and Terminology

In this section we define the basic concepts of this paper and set up standard terminology and notation. Important definitions will be preceded by a subtitle indicating the concept defined. The non-cooperative idea will be implicit, rather than explicit, below:

Finite Game

For us an *n-person game* will be a set of *n players*, or *positions*, each with an associated finite set of *pure strategies*; and corresponding to each player, *i*, a *payoff function*, p_i, which maps the set of all *n*-tuples of pure strategies into the real numbers. When we use the term *n-tuple* we shall always mean a set of *n* items, with each item associated with a different player.

Mixed Strategy, s_i

A *mixed strategy* of player *i* will be a collection of non-negative numbers which have unit sum and are in one to one correspondence with his pure strategies.

We write $s_i = \sum_a c_{i\alpha} \pi_{i\alpha}$ with $c_{i\alpha} \geqq 0$ and $\sum_\alpha c_{i\alpha} = 1$ to represent such a mixed strategy, where the $\pi_{i\alpha}$'s are the pure strategies of player *i*. We regard the s_i's as points in a simplex whose vertices are the $\pi_{i\alpha}$'s. This simplex may be regarded as a convex subset of a real vector space, giving us a natural process of linear combination for the mixed strategies.

We shall use the suffixes *i*, *j*, *k* for players and α, β, γ to indicate various pure strategies of a player. The symbols s_i, t_i, and r_i, etc. will indicate mixed strategies; $\pi_{i\alpha}$ will indicate the ith player's α^{th} pure strategy, etc.

Payoff Function, p_i

The payoff function, p_i, used in the definition of a finite game above, has a unique extension to the *n*-tuples of mixed strategies which is linear in the mixed strategy of each player [n-linear]. This extension we shall also denote by p_i, writing $p_i(s_1, s_2, \cdots, s_n)$.

We shall write $\mathfrak{s}$ or t to denote an n-tuple of mixed strategies and if $\mathfrak{s} = (s_1, s_2, \cdots, s_n)$ then $p_i(\mathfrak{s})$ shall mean $p_i(s_1, s_2, \cdots, s_n)$. Such an n-tuple, $\mathfrak{s}$, will also be regarded as a point in a vector space, the product space of the vector spaces containing the mixed strategies. And the set of all such n-tuples forms, of course, a convex polytope, the product of the simplices representing the mixed strategies.

For convenience we introduce the substitution notation $(\mathfrak{s}; t_i)$ to stand for $(s_1, s_2, \cdots, s_{i-1}, t_i, s_{i+1}, \cdots, s_n)$ where $\mathfrak{s} = (s_1, s_2, \cdots, s_n)$. **87** The effect of successive substitutions $((\mathfrak{s}; t_i); r_j)$ we indicate by $(\mathfrak{s}; t_i; r_j)$, etc.

Equilibrium Point

An n-tuple $\mathfrak{s}$ is an *equilibrium point* if and only if for every i

$$p_i(\mathfrak{s}) = \max_{\text{all } r_i\text{'s}} \left[p_i(\mathfrak{s}; r_i) \right]. \tag{1}$$

Thus an equilibrium point is an n-tuple $\mathfrak{s}$ such that each player's mixed strategy maximizes his payoff if the strategies of the others are held fixed. Thus each player's strategy is optimal against those of the others. We shall occasionally abbreviate equilibrium point by eq. pt.

We say that a mixed strategy s_i *uses* a pure strategy $\pi_{i\alpha}$ if $s_i = \sum_\beta c_{i\beta} \pi_{i\beta}$ and $c_{i\alpha} > 0$. If $\mathfrak{s} = (s_1, s_2, \cdots, s_n)$ and s_i uses $\pi_{i\alpha}$ we also say that $\mathfrak{s}$ uses $\pi_{i\alpha}$.

From the linearity of $p_i(s_1, \cdots, s_n)$ in s_i,

$$\max_{\text{all } r_i\text{'s}} \left[p_i(\mathfrak{s}; r_i) \right] = \max_\alpha \left[p_i(\mathfrak{s}; \pi_{i\alpha}) \right]. \tag{2}$$

We define $p_{i\alpha}(\mathfrak{s}) = p_i(\mathfrak{s}; \pi_{i\alpha})$. Then we obtain the following trivial necessary and sufficient condition for $\mathfrak{s}$ to be an equilibrium point:

$$p_i(\mathfrak{s}) = \max_\alpha p_{i\alpha}(\mathfrak{s}). \tag{3}$$

If $\mathfrak{s} = (s_1, s_2, \cdots, s_n)$ and $s_i = \sum_\alpha c_{i\alpha} \pi_{i\alpha}$ then $p_i(\mathfrak{s}) = \sum_\alpha c_{i\alpha} p_{i\alpha}(\mathfrak{s})$, consequently for (3) to hold we must have $c_{i\alpha} = 0$ whenever $p_{i\alpha}(\mathfrak{s}) < \max_\beta p_{i\beta}(\mathfrak{s})$, which is to say that $\mathfrak{s}$ does not use $\pi_{i\alpha}$ unless it is an optimal pure strategy for player i. So we write

$$\text{if } \pi_{i\alpha} \text{ is used in } \mathfrak{s} \text{ then } p_{i\alpha}(\mathfrak{s}) = \max_{\beta} p_{i\beta}(\mathfrak{s}) \qquad (4)$$

as another necessary and sufficient condition for an equilibrium point.

Since a criterion (3) for an eq. pt. can be expressed by the equating of n pairs of continuous functions on the space of n-tuples $\mathfrak{s}$ the eq. pts. obviously form a closed subset of this space. Actually, this subset is formed from a number of pieces of algebraic varieties, cut out by other algebraic varieties.

Existence of Equilibrium Points

A proof of this existence theorem based on Kakutani's generalized fixed point theorem was published in Proc. Nat. Acad. Sci. U.S.A., 36, pp. 48–49. The proof given here is a considerable improvement over that earlier version and is based directly on the Brouwer theorem. We proceed by constructing a continuous transformation T of the space of n-tuples such that the fixed points of T are the equilibrium points of the game.

THEOREM 1. *Every finite game has an equilibrium point.*

Proof. Let $\mathfrak{s}$ be an n-tuple of mixed strategies, $p_i(\mathfrak{s})$ the corresponding payoff to player i, and $p_{i\alpha}(\mathfrak{s})$ the payoff to player i if he changes to his α^{th} pure strategy $\pi_{i\alpha}$ and the others continue to use their respective mixed strategies from $\mathfrak{s}$. We now define a set of continuous functions of $\mathfrak{s}$ by

$$\varphi_{i\alpha}(\mathfrak{s}) = \max(0, p_{i\alpha}(\mathfrak{s}) - p_i(\mathfrak{s}))$$

and for each component s_i of $\mathfrak{s}$ we define a modification s_i' by

$$s_i' = \frac{s_i + \sum_\alpha \varphi_{i\alpha}(\mathfrak{s})\pi_{i\alpha}}{1 + \sum_\alpha \varphi_{i\alpha}(\mathfrak{s})},$$

calling $\mathfrak{s}'$ the n-tuple $(s_1', s_2', s_3' \cdots s_n')$.

We must now show that the fixed points of the mapping $T: \mathfrak{s} \to \mathfrak{s}'$ are the equilibrium points.

First consider any n-tuple $\mathfrak{s}$. In $\mathfrak{s}$ the ith player's mixed strategy s_i

will use certain of his pure strategies. Some one of these strategies, say $\pi_{i\alpha}$, must be "least profitable" so that $p_{i\alpha}(\mathfrak{s}) \leqq p_i(\mathfrak{s})$. This will make $\varphi_{i\alpha}(\mathfrak{s}) = 0$.

Now if this n-tuple $\mathfrak{s}$ happens to be fixed under T the proportion of $\pi_{i\alpha}$ used in s_i must not be decreased by T. Hence, for all β's, $\varphi_{i\beta}(\mathfrak{s})$ must be zero to prevent the denominator of the expression defining s'_i from exceeding 1.

Thus, if $\mathfrak{s}$ is fixed under T, for any i and β $\varphi_{i\beta}(\mathfrak{s}) = 0$. This means no player can improve his payoff by moving to a pure strategy $\pi_{i\beta}$. But this is just a criterion for an eq. pt. [see (2)]. **89**

Conversely, if $\mathfrak{s}$ is an eq. pt. it is immediate that all φ's vanish, making $\mathfrak{s}$ a fixed point under T.

Since the space of n-tuples is a cell the Brouwer fixed point theorem requires that T must have at least one fixed point $\mathfrak{s}$, which must be an equilibrium point.

Symmetries of Games

An *automorphism*, or *symmetry*, of a game will be a permutation of its pure strategies which satisfies certain conditions, given below.

If two strategies belong to a single player they must go into two strategies belonging to a single player. Thus if ϕ is the permutation of the pure strategies it induces a permutation ψ of the players.

Each n-tuple of pure strategies is therefore permuted into another n-tuple of pure strategies. We may call χ the induced permutation of these n-tuples. Let ξ denote an n-tuple of pure strategies and $p_i(\xi)$ the payoff to player i when the n-tuple ξ is employed. We require that if

$$j = i^\psi \quad \text{then} \quad p_j(\xi^\chi) = p_i(\xi)$$

which completes the definition of a symmetry.

The permutation ϕ has a unique linear extension to the mixed strategies. If

$$s_i = \sum_\alpha c_{i\alpha} \pi_{i\alpha} \quad \text{we define} \quad (s_i)^\phi = \sum_\alpha c_{i\alpha} (\pi_{i\alpha})^\phi.$$

The extension of ϕ to the mixed strategies clearly generates an extension of χ to the n-tuples of mixed strategies. We shall also denote this by χ.

We define a *symmetric n-tuple* $\mathfrak{s}$ of a game by $\mathfrak{s}^\chi = \mathfrak{s}$ for all χ's.

THEOREM 2. *Any finite game has a symmetric equilibrium point.*

Proof. First we note that $s_{i0} = \sum_\alpha \pi_{i\alpha} / \sum_\alpha 1$ has the property $(s_{i0})^\phi = s_{j0}$ where $j = i^\psi$, so that the n-tuple $\mathfrak{s}_0 = (s_{10}, s_{20}, \cdots, s_{n0})$ is fixed under any χ; hence any game has at least one symmetric n-tuple.

If $\mathfrak{s} = (s_1, \cdots, s_n)$ and $\mathfrak{t} = (t_1, \cdots, t_n)$ are symmetric then

$$\frac{\mathfrak{s} + \mathfrak{t}}{2} = \left(\frac{s_1 + t_1}{2}, \frac{s_2 + t_2}{2}, \cdots, \frac{s_n + t_n}{2} \right)$$

is also symmetric because $\mathfrak{s}^\chi = \mathfrak{s} \leftrightarrow s_j = (s_i)^\phi$, where $j = i^\psi$, hence

$$\frac{s_j + t_j}{2} = \frac{(s_i)^\phi + (t_i)^\phi}{2} = \left(\frac{s_i + t_i}{2} \right)^\phi,$$

hence

$$\left(\frac{\mathfrak{s} + \mathfrak{t}}{2} \right)^\chi = \frac{\mathfrak{s} + \mathfrak{t}}{2}.$$

This shows that the set of symmetric n-tuples is a convex subset of the space of n-tuples since it is obviously closed.

Now observe that the mapping $T : \mathfrak{s} \to \mathfrak{s}'$ used in the proof of the existence theorem was intrinsically defined. Therefore, if $\mathfrak{s}_2 = T\mathfrak{s}_1$ and χ is derived from an automorphism of the game we will have $\mathfrak{s}_2^\chi = T\mathfrak{s}_1^\chi$. If $\mathfrak{s}_1$ is symmetric $\mathfrak{s}_1^\chi = \mathfrak{s}_1$ and therefore $\mathfrak{s}_2^\chi = T\mathfrak{s}_1 = \mathfrak{s}_2$. Consequently this mapping maps the set of symmetric n-tuples into itself.

Since this set is a cell there must be a symmetric fixed point $\mathfrak{s}$ which must be a symmetric equilibrium point.

Solutions

We define here solutions, strong solutions, and sub-solutions. A non-cooperative game does not always have a solution, but when it does the solution is unique. Strong solutions are solutions with special properties. Sub-solutions always exist and have many of the properties of solutions, but lack uniqueness.

S_1 will denote a set of mixed strategies of player i and $\mathfrak{S}$ a set of n-tuples of mixed strategies.

Solvability

A game is *solvable* if its set, $\mathfrak{S}$, of equilibrium points satisfies the condition

$$(\mathfrak{t}; r_i) \in \mathfrak{S} \quad \text{and} \quad \mathfrak{s} \in \mathfrak{S} \rightarrow (\mathfrak{s}; r_i) \in \mathfrak{S} \quad \text{for all} \quad i\text{'s}. \quad (5)$$

This is called the *interchangeability* condition. The *solution* of a solvable game is its set, $\mathfrak{S}$, of equilibrium points.

Strong Solvability

A game is *strongly solvable* if it has a solution, $\mathfrak{S}$, such that for all i's

$$\mathfrak{s} \in \mathfrak{S} \quad \text{and} \quad p_i(\mathfrak{s}; r_i) = p_i(\mathfrak{s}) \rightarrow (\mathfrak{s}; r_i) \in \mathfrak{S}$$

and then $\mathfrak{S}$ is called a *strong solution*.

Equilibrium Strategies

In a solvable game let S_i be the set of all mixed strategies s_i such that for some $\mathfrak{t}$ the n-tuple $(\mathfrak{t}; s_i)$ is an equilibrium point. [s_i is the ith component of some equilibrium point.] We call S_i the set of *equilibrium strategies* of player i.

Sub-Solutions

If $\mathfrak{S}$ is a subset of the set of equilibrium points of a game and satisfies condition (1); and if $\mathfrak{S}$ is maximal relative to this property then we call $\mathfrak{S}$ a *sub-solution*.

For any sub-solution $\mathfrak{S}$ we define the ith *factor set*, S_i, as the set of all s_i's such that $\mathfrak{S}$ contains $(\mathfrak{t}; s_i)$ for some $\mathfrak{t}$.

Note that a sub-solution, when unique, is a solution; and its factor sets are the sets of equilibrium strategies.

THEOREM 3. *A sub-solution, $\mathfrak{S}$, is the set of all n-tuples $(s_1, s_2, \cdots, s_n)$ such that each $s_i \in S_i$ where S_i is the ith factor set of $\mathfrak{S}$. Geometrically, $\mathfrak{S}$ is the product of its factor sets.*

Proof. Consider such an n-tuple $(s_1, s_2, \cdots, s_n)$. By definition $\exists \mathfrak{t}_1$, $\mathfrak{t}_2, \cdots, \mathfrak{t}_n$ such that for each $i(\mathfrak{t}_i; s_i) \in \mathfrak{S}$. Using the condition

(5) $n - 1$ times we obtain successively $(t_1; s_1) \in \mathfrak{S}, (t_1; s_1; s_2) \in \mathfrak{S}, \cdots, (t_1; s_1; s_2; \cdots; s_n) \in \mathfrak{S}$ and the last is simply $(s_1, s_2, \cdots, s_n) \in \mathfrak{S}$, which we needed to show.

THEOREM 4. *The factor sets* $S_1, S_2, \cdots, S_n$ *of a sub-solution are closed and convex as subsets of the mixed strategy spaces.*

Proof. It suffices to show two things:

(a) if s_i and $s_i' \in S_i$ then $s_i^* = (s_i + s_i')/2 \in S_i$; (b) if $s_i^\#$ is a limit point of S_i then $s_i^\# \in S_i$.

Let $t \in \mathfrak{S}$. Then we have $p_j(t; s_i) \geqq p_j(t; s_i; r_j)$ and $p_j(t; s_i') \geqq p_j(t; s_i'; r_j)$ for any r_j, by using the criterion of (1) for an eq. pt. Adding these inequalities, using the linearity of $p_j(s_1, \cdots, s_n)$ in s_i, and dividing by 2, we get $p_j(t; s_i^*) \geqq p_j(t; s_i^*; r_j)$, since $s_i^* = (s_i + s_i')/2$. From this we know that $(t; s_i)$ is an eq. pt. for any $t \in \mathfrak{S}$. If the set of all such eq. pts. $(t; s_i^*)$ is added to $\mathfrak{S}$, the augmented set clearly satisfies condition (5), and since $\mathfrak{S}$ was to be maximal it follows that $s_i^* \in S_i$.

To attack (b), note that the n-tuple $(t; s_i^\#)$, where $t \in \mathfrak{S}$, will be a limit point of the set of n-tuples of the form $(t; s_i)$ where $s_i \in S_i$, since $s_i^\#$ is a limit point of S_i. But this set is a set of eq. pts. and hence any point in its closure is an eq. pt., since the set of all eq. pts. is closed. Therefore $(t; s_i^\#)$ is an eq. pt. and hence $s_i^\# \in S_i$ from the same argument as for s_i^*.

Values

Let $\mathfrak{S}$ be the set of equilibrium points of a game. We define

$$v_i^+ = \max_{\mathfrak{s} \in \mathfrak{S}}[\, p_i(\mathfrak{s})\,], \qquad v_i^- = \min_{\mathfrak{s} \in \mathfrak{S}}[\, p_i(\mathfrak{s})\,].$$

If $v_i^+ = v_i^-$ we write $v_i = v_i^+ = v_i^-$. v_i^+ is the *upper value* to player i of the game; v_i^- the *lower value*; and v_i the *value*, if it exists.

Values will obviously have to exist if there is but one equilibrium point.

One can define *associated values* for a sub-solution by restricting $\mathfrak{S}$ to the eq. pts. in the sub-solution and then using the same defining equations as above.

A two-person zero-sum game is always solvable in the sense defined above. The sets of equilibrium strategies S_1 and S_2 are simply the sets of "good" strategies. Such a game is not generally strongly solvable; strong solutions exist only when there is a "saddle point" in *pure* strategies.

Sample Examples

These are intended to illustrate the concepts defined in the paper and display special phenomena which occur in these games.

The first player has the roman letter strategies and the payoff to the left, etc.

Ex. 1

5	$a\alpha$	-3	Solution $\left(\frac{9}{16}a + \frac{7}{16}b, \frac{7}{17}a + \frac{10}{17}\beta\right)$
-4	$a\beta$	4	
-5	$b\alpha$	5	
3	$b\beta$	-4	$v_1 = \frac{-5}{17}, v_2 = +\frac{1}{2}$

Ex. 2

1	$a\alpha$	1	Strong Solution (b, β)
-10	$a\beta$	10	
10	$b\alpha$	-10	$v_1 = v_2 = 1$
-1	$b\beta$	-1	

Ex. 3

1	$a\alpha$	1	Unsolvable; equilibrium points (a, α), (b, β),
-10	$a\beta$	-10	and $\left(\frac{a}{2} + \frac{b}{2}, \frac{\alpha}{2} + \frac{\beta}{2}\right)$. The strategies in the last
-10	$b\alpha$	-10	case have maxi-min and mini-max
1	$b\beta$	1	properties.

Ex. 4

1	$a\alpha$	1	Strong Solution: all pairs of mixed
0	$a\beta$	1	strategies.
1	$b\alpha$	0	$v_1^+ = v_2^+ = 1, v_1^- = v_2^- = 0.$
0	$b\beta$	0	

Ex. 5

1	$a\alpha$	2	Unsolvable; eq. pts (a, α), (b, α) and
-1	$a\beta$	-4	$\left(\frac{1}{4}a + \frac{3}{4}b, \frac{3}{8}\alpha + \frac{5}{8}\beta\right)$. However, empirical tests
-4	$b\alpha$	-1	show a tendency toward (a, α).
2	$b\beta$	1	

Ex. 6

1	$a\alpha$	1	Eq. pts.: (a, α) and (b, β), with (b, β) an
0	$a\beta$	0	example of instability.
0	$b\alpha$	0	
0	$b\beta$	0	

Geometrical Form of Solutions

In the two-person zero-sum case it has been shown that the set of "good" strategies of a player is a convex polyhedral subset of his strategy space. We shall obtain the same result for a player's set of equilibrium strategies in any solvable game.

JOHN F. NASH, JR.

Non-Cooperative Games

THEOREM 5. *The sets $S_1, S_2, \cdots, S_n$ of equilibrium strategies in a solvable game are polyhedral convex subsets of the respective mixed strategy spaces.*

Proof. An n-tuple $\mathfrak{s}$ will be an equilibrium point if and only if for every i

$$p_i(\mathfrak{s}) = \max_\alpha p_{i\alpha}(\mathfrak{s}) \tag{6}$$

which is condition (3). An equivalent condition is for every i and α

$$p_i(\mathfrak{s}) - p_{i\alpha}(\mathfrak{s}) \geqq 0. \tag{7}$$

Let us now consider the form of the set S_j of equilibrium strategies, s_j, of player j. Let t be any equilibrium point, then $(t; s_j)$ will be an equilibrium point if and only if $s_j \in S_j$, from Theorem 2. We now apply conditions (2) to $(t; s_j)$, obtaining

$$s_j \in S_j \leftrightarrow \text{ for all } i, \alpha \qquad p_i(t; s_j) - p_i\alpha(t; s_j) \geqq 0. \tag{8}$$

Since p_i is n-linear and t is constant these are a set of linear inequalities of the form $F_{i\alpha}(s_j) \geqq 0$. Each such inequality is either satisfied for all s_j or for those lying on and to one side of some hyperplane passing through the strategy simplex. Therefore, the complete set [which is finite] of conditions will all be satisfied simultaneously on some convex polyhedral subset of player j's strategy simplex. [Intersection of half-spaces.]

As a corollary we may conclude that S_j is the convex closure of a finite set of mixed strategies [vertices].

Dominance and Contradiction Methods

We say that s_i' dominates s_i if $p_i(t; s_i') > p_i(t; s_i)$ for every t.

This amounts to saying that s_i' gives player i a higher payoff than s_i no matter what the strategies of the other players are. To see whether a strategy s_i' dominates s_i it suffices to consider only pure strategies for the other players because of the n-linearity of p_i.

It is obvious from the definitions that *no equilibrium point can involve a dominated strategy s_i.*

The domination of one mixed strategy by another will always entail other dominations. For suppose s_i' dominates s_i and t_i uses all of the pure strategies which have a higher coefficient in s_i than in s_i'. Then for a small enough ρ

$$t_i' = t_i + \rho(s_i' - s_i)$$

is a mixed strategy; and t_i dominates t_i' by linearity.

One can prove a few properties of the set of undominated strategies. It is simply connected and is formed by the union of some collection of faces of the strategy simplex.

The information obtained by discovering dominances for one player may be of relevance to the others, insofar as the elimination of classes of mixed strategies as possible components of an equilibrium point is concerned. For the t's whose components are all undominated are all that need be considered and thus eliminating some of the strategies of one player may make possible the elimination of a new class of strategies for another player.

Another procedure which may be used in locating equilibrium points is the contradiction-type analysis. Here one assumes that an equilibrium point exists having component strategies lying within certain regions of the strategy spaces and proceeds to deduce further conditions which must be satisfied if the hypothesis is true. This sort of reasoning may be carried through several stages to eventually obtain a contradiction indicating that there is no equilibrium point satisfying the initial hypothesis.

A Three-Man Poker Game

As an example of the application of our theory to a more or less realistic case we include the simplified poker game given below. The rules are as follows:

(a) The deck is large, with equally many *high* and *low* cards, and a hand consists of one card.
(b) Two chips are used to ante, open, or call.
(c) The players play in rotation and the game ends after all have passed or after one player has opened and the others have had a chance to call.
(d) If no one bets the antes are retrieved.
(e) Otherwise the pot is divided equally among the highest hands which have bet.

We find it more satisfactory to treat the game in terms of quantities we call "behavior parameters" than in the normal form of *Theory of Games and Economic Behavior*. In the normal form representation two mixed strategies of a player may be equivalent in the sense that each makes the individual choose each available course of action in each particular situation requiring action on his part with the same frequency. That is, they represent the same behavior pattern on the part of the individual.

Behavior parameters give the probabilities of taking each of the various possible actions in each of the various possible situations which may arise. Thus they describe behavior patterns.

In terms of behavior parameters the strategies of the players may be represented as follows, assuming that since there is no point in passing with a *high* card at one's last opportunity to bet that this will not be done. The greek letters are the probabilities of the various acts.

	First Moves	Second Moves
I	α Open on *high* β Open on *low*	κ Call III on *low* λ Call II on *low* μ Call II and III on *low*
II	γ Call I on *low* δ Open on *high* ϵ Open on *low*	ν Call III on *low* ξ Call III and I on *low*
III	ζ Call I and II on *low* η Open on *low* θ Call I on *low* ι Call II on *low*	Player III never gets a second move

We locate all possible equilibrium points by first showing that most of the greek parameters must vanish. By dominance mainly with a little contradiction-type analysis β is eliminated and with it go γ, ζ, and θ by dominance. Then contradictions eliminate μ, ξ, ι, λ, κ, and ν in that order. This leaves us with α, δ, ϵ, and η. Contradiction analysis shows that none of these can be zero or one and thus we obtain a system of simultaneous algebraic equations. The equations happen to have but one solution with the variables in the range (0, 1). We get

$$\alpha = \frac{21 - \sqrt{321}}{10}, \quad \eta = \frac{5\alpha + 1}{4}, \quad \delta = \frac{5 - 2\alpha}{5 + \alpha}, \quad \epsilon = \frac{4\alpha - 1}{\alpha + 5}.$$

These yields $\alpha = .308$, $\eta = .635$, $\delta = .826$, and $\epsilon = .044$. Since there is only one equilibrium point the game has values; these are

$$v_1 = -.147 = -\frac{(1 + 17\alpha)}{8(5 + \alpha)}, \quad v_2 = -.096 = \frac{1 - 2\alpha}{4},$$

and

$$v_3 = .243 = \frac{79}{40}\left(\frac{1 - \alpha}{5 + \alpha}\right).$$

A more complete investigation of this poker game is published in Annals of Mathematics Study No. 24, *Contributions to the Theory of Games*. There the solution is studied as the ratio of ante to bet varies, and the potentialities of coalitions are investigated.

Applications

The study of *n*-person games for which the accepted ethics of fair play imply non-cooperative playing is, of course, an obvious direction in which to apply this theory. And poker is the most obvious target. The analysis of a more realistic poker game than our very simple model should be quite an interesting affair.

The complexity of the mathematical work needed for a complete investigation increases rather rapidly, however, with increasing complexity of the game; so that analysis of a game much more complex than the example given here might only be feasible using approximate computational methods.

A less obvious type of application is to the study of cooperative games. By a cooperative game we mean a situation involving a set of players, pure strategies, and payoffs as usual; but with the assumption that the players can and will collaborate as they do in the von Neumann and Morgenstern theory. This means the players may communicate and form coalitions which will be enforced by an umpire. It is unnecessarily restrictive, however, to assume any transferability or even comparability of the payoffs [which should be in utility units] to different players. Any desired transferability can be put into the game itself instead of assuming it possible in the extra-game collaboration.

The writer has developed a "dynamical" approach to the study of cooperative games based upon reduction to non-cooperative form. One proceeds by constructing a model of the pre-play negotiation so that the steps of negotiation become moves in a larger non-cooperative game [which will have an infinity of pure strategies] describing the total situation.

This larger game is then treated in terms of the theory of this paper [extended to infinite games] and if values are obtained they are taken as the values of the cooperative game. Thus the problem of analyzing a cooperative game becomes the problem of obtaining a suitable, and convincing, non-cooperative model for the negotiation.

The writer has, by such a treatment, obtained values for all finite two-person cooperative games, and some special n-person games.

Acknowledgments

Drs. Tucker, Gale, and Kuhn gave valuable criticism and suggestions for improving the exposition of the material in this paper. David Gale suggested the investigation of symmetric games. The solution of the Poker model was a joint project undertaken by Lloyd S. Shapley and the author. Finally, the author was sustained financially by the Atomic Energy Commission in the period 1949–50 during which this work was done.

Bibliography

1. von Neumann, Morgenstern, Theory of Games and Economic Behavior, Princeton University Press, 1944.
2. J. F. Nash, Jr., *Equilibrium Points in n-Person Games,* Proc. Nat. Acad. Sci. U.S.A. 36 (1950) 48–49.
3. J. F. Nash, L. S. Shapley, A Simple Three-Person Poker Game, Annals of Mathematics Study No. 24, Princeton University Press, 1950.
4. John Nash, *Two Person Cooperative Games,* to appear in Econometrica.
5. H. W. Kuhn, *Extensive Games,* Proc. Nat. Acad. Sci. U.S.A. 36 (1950) 570–576.

CHAPTER 8

JOHN F. NASH, JR.

Two-Person Cooperative Games

In this paper, the author extends his previous treatment of "The Bargaining Problem" to a wider class of situations in which threats can play a role. A new approach is introduced involving the elaboration of the threat concept.

Introduction

The theory presented here was developed to treat economic (or other) situations involving two individuals whose interests are neither completely opposed nor completely coincident. The word cooperative is used because the two individuals are supposed to be able to discuss the situation and agree on a rational joint plan of action, an agreement that should be assumed to be enforceable.

It is conventional to call these situations "games" when they are being studied from an abstract mathematical viewpoint. Here the original situation is reduced to a mathematical description, or model. In the

This paper was written with the support of The RAND Corporation. It appeared in an earlier form as RAND P-172, August 9, 1950.

abstract "game" formulation only the minimum quantity of information necessary for the solution is retained. What the actual alternative courses of action are among which the individuals must choose is not regarded as essential information. These alternatives are treated as abstract objects without special qualities and are called "strategies." Only the attitudes (like or dislike) of the two individuals towards the ultimate results of the use of the various possible opposing pairs of strategies are considered; but this information must be well utilized and must be expressed quantitatively.

The theory of von Neumann and Morgenstern applies to some of the games considered here. Their assumption that it is possible for the players to make "side-payments" in a commodity for which each individual (player) has a linear utility narrows the range of their theory's applicability. In this paper there is no assumption about side-payments. If the situation permits side-payments then this simply affects the set of possible final outcomes of the game; side-payments are treated just like any other activity that may take place in the actual playing of the game—no special consideration is necessary. The von Neumann and Morgenstern approach also differs by giving a much less determinate solution. Their approach leaves the final situation only determined up to a side-payment. The side-payment is generally not determined but is restricted to lie in a certain range.

An earlier paper by the author [3] treated a class of games which are in one sense the diametrical opposites of the cooperative games. A game is non-cooperative if it is impossible for the players to communicate or collaborate in any way. The non-cooperative theory applies without change to any number of players, but the cooperative case, which is analyzed in this paper, has only been worked out for two players.

We give two independent derivations of our solution of the two-person cooperative game. In the first, the cooperative game is reduced to a non-cooperative game. To do this, one makes the players' steps of negotiation in the cooperative game become moves in the non-cooperative model. Of course, one cannot represent all possible bargaining devices as moves in the non-cooperative game. The negotiation process must be formalized and restricted, but in such a way that each participant is still able to utilize all the essential strengths of his position.

The second approach is by the axiomatic method. One states as axioms several properties that it would seem natural for the solution to have and then one discovers that the axioms actually determine the solution uniquely. The two approaches to the problem, via the negotiation model or via the axioms, are complementary; each helps to justify and clarify the other.

The Formal Representation of the Game

Each of the players (one and two) has a compact convex metrizable space S_i of mixed strategies s_i (those readers who are unacquainted with the mathematical technicalities will find that they can manage quite well by ignoring them). These mixed strategies represent the courses of action player i can take independently of the other player. They may involve deliberate decisions to randomize, to decide between alternative possibilities by using a randomizing process involving specified probabilities. This randomizing is an essential ingredient in the concept of a mixed strategy. By beginning with a space of mixed strategies instead of talking about a sequence of moves, etc., we presuppose a reduction of the strategic potentialities of each player to the normal form [4].

The possible joint courses of action by the players would form a similar space. But the only important thing is the set of those pairs (u_1, u_2) of utilities which can be realized by the players if they cooperate. We call this set B and it should be a compact convex set in the (u_1, u_2) plane.

For each pair (s_1, s_2) of strategies from S_1 and S_2, there will be the utility to each player of a situation where these strategies are to be employed or carried out. These utilities (pay-offs in game theoretic usage) are denoted by $p_1(s_1, s_2)$ and $p_2(s_1, s_2)$. Each p_i is a linear function of s_1 and of s_2, although it cannot be expected to depend linearly on the two varying simultaneously; in other words, p_i is a bilinear function of s_1 and s_2. Basically this linearity is a consequence of the type of utility we assume for the players; it is thoroughly discussed in an early chapter of von Neumann and Morgenstern [4].

And of course each point in the (u_1, u_2) plane of the form $[p_1(s_1, s_2), p_2(s_1, s_2)]$ must be a point in B because every pair (s_1, s_2) of inde-

JOHN F. NASH, JR.

Two-Person Cooperative Games

pendent strategies corresponds to a joint policy (probably an inefficient one). This remark completes the formal, or mathematical, description of the game.

The Negotiation Model

To explain and justify the negotiation model used to obtain the solution we must say more about the general assumptions about the situation facing the two individuals, or, what it amounts to, about the conditions under which the game is to be played.

Each player is assumed fully informed on the structure of the game *and* on the utility function of his co-player (of course he also knows his own utility function). (This statement must not be construed as inconsistent with the indeterminacy of utility functions up to transformations of the form $u' = au + b, a > 0$). These information assumptions should be noted, for they are not generally perfectly fulfilled in actual situations. The same goes for the further assumption we need that the players are intelligent, rational individuals.

A common device in negotiation is the threat. The threat concept is really basic in the theory developed here. It turns out that the solution of the game not only gives what should be the utility of the situation to each player, but also tells the players what threats they should use in negotiating.

If one considers the process of making a threat, one sees that its elements are as follows: A threatens B by convincing B that if B does not act in compliance with A's demands, then A will follow a certain policy T. Supposing A and B to be rational beings, it is essential for the success of the threat that A be *compelled* to carry out his threat T if B fails to comply. Otherwise it will have little meaning. For, in general, to execute the threat will not be something A would want to do, just of itself.

The point of this discussion is that we must assume there is an adequate mechanism for forcing the players to stick to their threats and demands once made; and one to enforce the bargain, once agreed. Thus we need a sort of umpire, who will enforce contracts or commitments.

And in order that the description of the game be complete, we must suppose that the players have no prior commitments that might affect the game. We must be able to think of them as completely free agents.

The Formal Negotiation Model

Stage one: Each player (i) chooses a mixed strategy t_i, which he will be forced to use if the two cannot come to an agreement, that is, if their demands are incompatible. This strategy t_i is player i's threat.

Stage two: The players inform each other of their threats.

Stage three: In this stage the players act independently and without communication. The assumption of independent action is essential here, whereas no special assumptions of this type are needed in Stage one, as it turns out. In Stage three, each player decides upon his demand d_i, which is a point on his utility scale. The idea is that player i will not cooperate unless the mode of cooperation has at least the utility d_i to him.

Stage four: The pay-offs are now determined. If there is a point (u_1, u_2) in B such that $u_1 \geq d_1$ and $u_2 \geq d_2$, then the pay-off to each player i is d_i. That is, if the demands can be simultaneously satisfied, then each player gets what he demanded. Otherwise, the pay-off to player i is $p_i(t_1, t_2)$; i.e., the threat must be executed.

The choice of the pay-off function in the case of compatible demands may seem unreasonable, but it has its advantages. It cannot be accused of contributing a bias to the final solution and it gives the players a strong incentive to increase their demands as much as is possible without losing compatibility. But it can be embarrassingly accused of picking points that are not in the set B. Effectively, we have enlarged B to a set including all utility pairs dominated (weakly; $u'_1 \leq u_1$, $u'_2 \leq u_2$) by a pair in B.

What we have is actually a two move game. Stages two and four do not involve any decisions by the players. The second move choices are made with full information about what was done in the first move. Therefore, the game consisting of the second move alone may be considered separately (it is a game with a variable pay-off function determined by the choices made at the first move). The effect of the choice of threats on this game is to determine the pay-offs if the players do not cooperate.

Let N be the point $[p_1(t_1, t_2), p_2(t_1, t_2)]$ in B. This point N represents the effect of the use of the threats. Let u_{1N} and u_{2N} abbreviate the coordinates of N. If we introduce a function $g(d_1, d_2)$ which is $+1$ for compatible demands and 0 for incompatible demands, then we can represent the pay-offs as follows:

$$\text{to player one} \qquad d_1g + u_{1N}(1 - g),$$
$$\text{to player two} \qquad d_2g + u_{2N}(1 - g).$$

The demand game defined by these pay-off functions will generally have an infinite number of inequivalent equilibrium points [3]. Every pair of demands which graphs as a point on the upper-right boundary of B and which is neither lower nor to the left of N will form an equilibrium point. Thus the equilibrium points do not lead us immediately to a solution of the game. But if we discriminate between them by studying their relative stabilities we can escape from this troublesome non-uniqueness.

To do this we "smooth" the game to obtain a continuous pay-off function and then study the limiting behavior of the equilibrium points of the smoothed game as the amount of smoothing approaches zero.

A certain general class of natural smoothing methods will be considered here. This class is broader than one might at first think, for many other methods that superficially seem different are actually equivalent.

To smooth the game we approximate the discontinuous function g by a continuous function h, which has a value near to g's value except at the points near the boundary of B, where g is discontinuous. The function $h(d_1, d_2)$ should be thought of as representing the probability of compatibility of the demands d_1 and d_2. It can be thought of as representing uncertainties in the information structure of the game, the utility scales, etc. For convenience, let us assume that $h = 1$ on B and that h tapers off very rapidly towards zero as (d_1, d_2) moves away from B, without ever actually reaching zero. Another simplification can be had by assuming the utility functions properly transformed so that $u_{1N} = u_{2N} = 0$. Then we can write the pay-off functions for the smoothed game as $P_1 = d_1h$, $P_2 = d_2h$. For the original game h is replaced by g.

A pair of demands (d_1, d_2) viewed as a pair of pure strategies in the demand game, will be an equilibrium point if p_1, which is $d_1 h$, is maximized here for constant d_2 and if $p_2 = d_2 h$ is maximized for constant d_1. Now suppose (d_1, d_2) is a point where $d_1 d_2 h$ is maximized over the whole region in which d_1 and d_2 are positive. Then $d_1 h$ and $d_2 h$ will be maximized for constant d_2 and d_1, respectively, and (d_1, d_2) must be an equilibrium point.

If the function h decreases with increasing distance from B in a wavy or irregular way, there may be more equilibrium points and perhaps even more points where $d_1 d_2 h$ is a maximum. But if h varies regularly there will be only one equilibrium point coinciding with a unique maximum of $d_1 d_2 h$. However, we do not need to appeal to a regular h to justify the solution.

Let P be any point where $d_1 d_2 h$ or, what is the same thing, $u_1 u_2 h$ is maximized as above described and let ρ be the maximum of $u_1 u_2$ on the part of B lying in the region $u_1 \geq 0$, $u_2 \geq 0$. The value of $u_1 u_2$ at P must be at least ρ, since $0 \leq h \leq 1$ and since $h = 1$ on B. Figure 1 illustrates this situation. In it, Q is the point where $u_1 u_2$ is maximized on B (in the first quadrant about N) and $\alpha\beta$ is the hyperbola $u_1 u_2 = \rho$, which touches B at Q.

The important observation is that P must lie above $\alpha\beta$ but still be near enough to B for h to nearly equal 1. And as less and less smoothing is used, h will decrease more and more rapidly on moving away from B; hence any maximum point P of $u_1 u_2 h$ will have to be nearer and nearer to B. In the limit all such points must approach Q, the only contact point of B and the area above $\alpha\beta$. Thus Q is a necessary limit of equilibrium points, and Q is the only one.

We take Q for the solution of the demand game, characterized as the *only necessary limit of the equilibrium points of smoothed games*. The values of u_1 and u_2 at Q will be taken as the values of the demand game and as the optimal demands.

The discussion above implicitly assumed that B contained points where $u_1 > 0$, $u_2 > 0$ (after the normalization which made $u_1 = u_2 = 0$ at N). The other cases can be treated more simply without resource to a smoothing process. In these "degenerate cases" there is only one point of B which dominates the point N and is not itself dominated by some other point of B. (A point (u_1, u_2) is dominated by another point (u'_1, u'_2) if $u'_1 \geq u_1$ and $u'_2 \geq u_2$) (see Figure 1). This gives us the natural solution in these cases.

One should note that the solution point Q of the demand game varies as a continuous function of the threat point N. Also there is a helpful geometrical characterization of the way Q depends on N. The solution point Q is the contact point with B of a hyperbola whose asymptotes are the vertical and horizontal lines through N. Let T be the tangent at Q to this hyperbola (see Figure 2).

If linear transformations are applied to the utility functions, N can be made the origin and Q the point $(1, 1)$. Now T will have slope -1 and the line NQ will have slope $+1$. The essential point is that slope T = minus slope NQ, because this is a property that is not destroyed by linear transformations of the utilities. T will be a support line for the set B (that is, a line such that all points of B are either on the lower left side of T or are on T itself; for a proof, see reference [2] where the same situation arises).

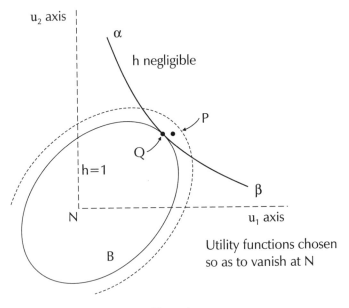

Figure I.

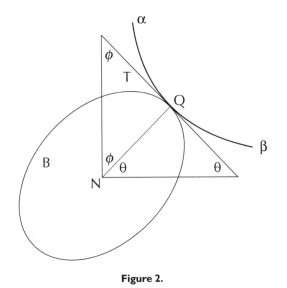

Figure 2.

We can now state the criterion: if NQ has positive slope and a support line T for B passes through Q with a slope equal but opposite to NQ's slope, then Q is the solution point for the threat point N. If NQ is horizontal/vertical and is itself a support line for B and if Q is the rightmost/uppermost of the points common to B and NQ, then again Q is the solution point for N (see Figure 3), and one of these cases must hold if Q is N's solution point; the criterion is a necessary and sufficient one.

Any support line of B with a contact point Q on the upper-right boundary of B determines a complementary line through Q with equal but opposite slope. All points on the line segment in which this complementary line intersects B are points which, as threat points, would have Q as corresponding solution point. The class of all these line segments is a ruling of B by line segments which intersect, if at all, only on the upper-right boundary of B. Given a threat point N, its solution point is the upper-right end of the segment passing through it (unless perhaps N is on more than one ruling and hence is on the upper-right boundary and is its own solution point).

We can now analyze the threat game, the game formed by the first move and with pay-off function determined by the solution of

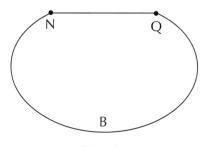

Figure 3.

the demand game. This pay-off is determined by the location of N, specifically by the ruling on which N falls. A ruling that is higher (or farther left) is more favorable to player two (let us definitely think of u_2 as measured on the vertical axis of the utility plane) and less favorable to player one.

Now if one player's threat is held fixed, say player one's at t_1, then the position of N is a function of the other player's threat, t_2. The coordinates of N, $p_1(t_1, t_2)$ and $p_2(t_1, t_2)$ are linear functions of t_2. Hence the transformation, t_2 goes into N, defined by this situation is a linear transformation of the space S_2 of player two's threats into B. That part of the image of S_2 that falls on the most favorable (for player two) ruling will contain the images of the threats that would be best as replies to player one's fixed particular threat t_1. And this set of best replies must be a convex compact subset of S_2 because of the linearity and continuity of the transformation of S_2 into B.

The continuity of N as a function of t_1 and t_2 and the continuity of Q as a function of N insure that the pay-off function defined for the threat game by solving the demand game is a continuous function of the threats. And this suffices to make each player's set of best replies what is called an upper semi-continuous function of the threat being replied to. Now consider any pair of threats (t_1, t_2). For each threat of the pair the other player has a set of best replies. Let $R(t_1, t_2)$ be the set of all pairs which contain one threat from each of the two sets of replies. R will be an upper semi-continuous function of (t_1, t_2) in the space of opposed pairs of threats, and $R(t_1, t_2)$ will always be a convex set in this space, $S_1 \times S_2$.

We are now ready to use the Kakutani fixed point theorem, as generalized by Karlin [1, pp. 159–160]. This theorem tells us that there is some pair (t_{10}, t_{20}) that is contained in its set $R(t_{10}, t_{20})$, which amounts to saying that each threat is a best reply to the other. Thus we have obtained an equilibrium point in the threat game. It is worth noting that this equilibrium point is formed by pure strategies in the threat game (a mixed strategy here would involve randomization over several threats).

The pair (t_{10}, t_{20}) also has minimax and maximin properties. Since the final pay-off in the game is determined by the position of Q on the upper-right boundary of B, which is a negatively sloping curve, each player's pay-off is a monotone decreasing function of the other's. So if player one sticks to t_{10}, player two cannot make one worse off than he does by using t_{20} without improving his own position, and he can't do this because (t_{10}, t_{20}) is an equilibrium point [3]. Thus t_{10} *assures* player one the equilibrium pay-off and t_{20} accomplishes the same for player two.

The threat game is now revealed to be very much like a zero-sum game, and one can readily see that if one player were to choose his threat first and inform the other, rather than their simultaneously choosing threats, this would not make any difference because there is a "saddle-point" in pure strategies. It is rather different with the demand game. The right to make the first demand would be quite valuable, so the simultaneity here is essential.

To summarize, we have now solved the negotiation model, found the values of the game to the two players, and shown that there are optimal threats and optimal demands (the optimal demands are the values).

The Axiomatic Approach

Rather than solve the two-person cooperative game by analyzing the bargaining process, one can attack the problem axiomatically by stating general properties that "any reasonable solution" should possess. By specifying enough such properties one excludes all but one solution.

The axioms below lead to the same solution that the negotiation model gave us; yet the concepts of demand or threat do not appear in them. Their concern is solely with the relationship between the solution (interpreted here as the value) of the game and the basic spaces and functions which give the mathematical description of the game.

It is rather significant that this quite different approach yields the same solution. This indicates that the solution is appropriate for a wider variety of situations than those which satisfy the assumptions we made in the approach via the model.

The notation used below is the same as before, except for a few additions. A triad (S_1, S_2, B) stands for a game and $v_1(S_1, S_2, B)$ and $v_2(S_1, S_2, B)$ are its values to the two players. Of course the triadic representation (S_1, S_2, B), leaves implicit the pay-off functions $p_1(s_1, s_2)$ and $p_2(s_1, s_2)$ which must be given to determine a game.

AXIOM I. For each game (S_1, S_2, B) there is a unique solution (v_1, v_2) which is a point in B.

AXIOM II. If (u_1, u_2) is in B and $u_1 \geq v_1$ and $u_2 \geq v_2$ then $(u_1, u_2) = (v_1, v_2)$; that is, the solution is not weakly dominated by any point in B except itself.

AXIOM III. Order preserving linear transformations of the utilities ($u_1' = a_1 u_1 + b_1$, $u_2' = a_2 u_2 + b_2$ with a_1 and a_2 positive) do not change the solution. It is understood that the numerical values will be changed by the direct action of the utility transformations, but the relative position of (v_1, v_2) in B should stay the same.

AXIOM IV. The solution does not depend on which player is called player one. In other words, it is a symmetrical function of the game.

AXIOM V. If a game is changed by restricting the set B of attainable pairs of utilities and the new set B' still contains the solution point of the original game, then this point will also be the solution point of the new game. Of course the new set B' must still contain all points of the form $[p_1(s_1, s_2), p_2(s_1, s_2)]$, where s_1 and s_2 range over S_1 and S_2, to make (S_1, S_2, B') a legitimate game.

AXIOM VI. A restriction of the set of strategies available to a player cannot increase the value to him of the game. Symbolically, if S_1' is contained in S_1, then $v_1(S_1', S_2, B) \leq v_1(S_1, S_2, B)$.

AXIOM VII. There is some way of restricting both players to single strategies without increasing the value to player one of the game. In symbols, there exist s_1 and s_2 such that $v_1(s_1, s_2, B) \leq v_1(S_1, S_2, B)$. Similarly, there is a way to do the same for player two.

There is little need to comment on Axiom I; it is just a statement on the type of solution desired. Axiom II expresses the idea that the players should succeed in cooperating with optimal efficiency. The principle of noncomparability of utilities is expressed in Axiom III. Each player's utility function is regarded as determined only up to order preserving linear transformations. This indeterminacy is a natural consequence of the definition of utility [4, chapter 1, part 3]. To reject Axiom III is to assume that some additional factor besides each individual's relative preferences for alternatives is considered to make the utility functions more determinate and to assume that this factor is significant in determining the outcome of the game.

The symmetry axiom, Axiom IV, says that the only significant (in determining the value of the game) differences between the players are those which are included in the mathematical description of the game, which includes their different sets of strategies and utility functions. One may think of Axiom IV as requiring the players to be intelligent and rational beings. But we think it is a mistake to regard this as expressing "equal bargaining ability" of the players, in spite of a statement to this effect in "The Bargaining Problem" [2]. With people who are sufficiently intelligent and rational there should not be any question of "bargaining ability," a term which suggests something like skill in duping the other fellow. The usual haggling process is based on imperfect information, the hagglers trying to propagandize each other into misconceptions of the utilities involved. Our assumption of complete information makes such an attempt meaningless.

It is probably harder to give a good plausibility argument for

Axiom V than for any of the others. There is some discussion of it in "The Bargaining Problem" [2]. This axiom is equivalent to an axiom of "localization" of the dependence of the solution point on the shape of the set B. The location of the solution point on the upper-right boundary of B is determined only by the shape of any small segment of the boundary that extends to both sides of it. It does not depend on the rest of the boundary curve.

Thus there is no "action at a distance" in the influence of the shape of B on the location of the solution point. Thinking in terms of bargaining, it is as if a proposed deal is to compete with small modifications of itself and that ultimately the negotiation will be understood to be restricted to a narrow range of alternative deals and to be unconcerned with more remote alternatives.

The last two axioms are the only ones that are primarily concerned with the strategy spaces S_1 and S_2, and the only ones that are really new. The other axioms are simply appropriate modifications of the axioms used in "The Bargaining Problem." Axiom VI says that a player's position in the game is not improved by restricting the class of threats available to him. This is surely reasonable.

The need for Axiom VII is not immediately obvious. Its effect is to remove the possibility that the value to a player of his space of threats should be dependent on collective or mutual reinforcement properties of the threats. The way Axiom VII is used in the demonstration of the adequacy of the axioms probably reveals its real content better than any heuristic discussion we might give here.

We can shortcut some of the arguments needed to show that the axioms accomplish their purpose and characterize the same solution we obtained with the model by appealing to the results of "The Bargaining Problem." We first consider games where each player has but one possible threat. Such a game is essentially a "bargaining problem," and for that sort of game our Axioms I, II, III, IV, and V are the same as the axioms of "The Bargaining Problem."

This determines the solution in the case where each player has but one strategy available. It must be the same solution obtained in "The Bargaining Problem," which was the same as the solution we got for the demand game (which is played after each player has chosen a threat) in the preceding approach. This solution is characterized by the maximization of the product, $[v_1 - p_1(t_1, t_2)][v_2 - p_2(t_1, t_2)]$, of the differences between the values of the game and the utilities of the situation where the players do not cooperate.

However, we are obliged to remark that the situation to be treated here is more general than that in "The Bargaining Problem" because it was assumed in that paper that there was some way for the players to cooperate with mutual benefit. Here it may be the case that only one, or neither, of the players can actually gain by cooperation. To show that the axioms handle this case seems to require a more complicated argument using Axioms VI and VII. But this is a minor point and we shall not include that argument, which is long out of proportion to its significance.

The primary function of Axioms VI and VII is to enable us to reduce the problem of games where each player may have a non-trivial space of strategies (threats) to the case we have just dealt with, where each has but one possible threat. Suppose player one is restricted to a strategy t_{10} which would be an optimal threat in the threat game discussed before in the non-axiomatic approach. Then from Axiom VI, we have

$$v_1(t_{10}, S_2, B) \leq v_1(S_1, S_2, B).$$

Now we apply Axiom VII to restrict S_2 to a single strategy (S_1 is already restricted) without increasing the value of the game to player one. Let t_2^* stand for the single strategy that S_2 is restricted to, then

$$v_1(t_{10}, t_2^*, B) \leq v_1(t_{10}, S_2, B).$$

Now we know that the value of a game where each player has but one threat is the same value obtained in the first part of this paper. Hence we know that against the threat t_{10} there is no better threat for player two, and no threat more unfavorable for player one, that t_{20} (i.e., an optimal threat for player two). So we may write

$$v_1(t_{10}, t_{20}, B) \leq v_1(t_{10}, t_2^*, B).$$

Combining the three inequalities we have

$$v_1(t_{10}, t_{20}, B) \leq v_1(S_1, S_2, B).$$

J O H N   F .   N A S H ,   J R .

Two-Person Cooperative Games

Similarly, we have

$$v_2(t_{10}, t_{20}, B) \leq v_2(S_1, S_2, B).$$

And now we observe, by Axiom II, that the last two inequalities may be replaced by equalities, because $v_1(t_{10}, t_{20}, B)$ and $v_2(t_{10}, t_{20}, B)$ are the coordinates of a point on the upper-right boundary of B. Thus the axiomatic approach gives the same values as the other approach.

114 References

1. Kuhn, H. W., and A. W. Tucker, eds., *Contributions to the Theory of Games (Annals of Mathematics Study No. 24)*, Princeton: Princeton University Press, 1950, 201 pp.
2. Nash, John, "The Bargaining Problem," *Econometrica*, Vol. 18, April, 1950, pp. 155–62.
3. Nash, John, "Non-Cooperative Games," *Annals of Mathematics*, Vol. 54, September 1951, pp. 286–95.
4. von Neumann, J., and O. Morgenstern, *Theory of Games and Economic Behavior*, 2nd edition, Princeton: Princeton University Press, 1947, 641 pp.

Editor's Introduction to Chapter 9

Modern computing began with the proposal, made by Alan Turing and John von Neumann, among others, to prepare a set of instructions for a single processor that would execute the instructions one after another. Such a set of instructions is called a "computer program" and a single processor executes the instructions "serially." If a number of processors are available, then several operations can be performed simultaneously, or in "parallel." Thus, if we wish to compute a sum of 100 terms serially, in which each term is the product of two numbers, $a(1)b(1) + a(2)b(2) + \ldots + a(100)b(100)$, we will need 100 multiplications and 99 additions done one after another. If we have 100 processors available, then the 100 multiplications can be done in one step and, by grouping terms, the additions can be done in seven successive steps (an exercise for the reader). Thus 199 serial steps have been done in eight steps, a dramatic reduction in computing time. The paper that follows is a proposal by Nash for the architecture of such a parallel computer.

It is difficult to reconstruct the early history of parallel computing. A very frequently quoted timeline of the early history (available on the Web at http://ei.cs.vt.edu/~history/Parallel.html) starts with the construction of the IBM 704, designed by Gene Amdahl in 1955; Amdahl was later to argue against the idea of parallel computing. This history continues in 1956 with the IBM 7030 project known as STRETCH, the LARC project for Lawrence Livermore National Laboratory, and the Atlas project in the United Kingdom.

The RAND memorandum that follows is dated August 27, 1954, well before these attempts to accomplish practical parallel computation on an actual machine. We do not know whether Nash's proposal was taken seriously by the computing group at RAND. It, however, stands as evidence of a rare imagination.

[H.W.K.]

CHAPTER 9

JOHN F. NASH, JR.

Parallel Control

This memorandum concerns some ideas for new designs of the control system in high-speed digital computers. The ideas are yet in an immature and rather unspecific form, but this is a subject that deserves some attention and thought for the future.

Indeed, the idea is more or less futuristic, and is more appropriate for the "electronic brains" of the future than for the computers now used, or under construction, or even planned. The basic idea is simple. Instead of having a single control unit sequencing the operations of the machine in series (except for certain subsidiary operations as certain input and output functions) as is now done, the idea is to decentralize control with several different control units capable of directing various simultaneous operations and interrelating them when appropriate.

Let us consider some of the advantages of this sort of development of computing (and data processing) machines.

1. *Speed of computation.* This would apply mainly to problems admitting computations in parallel. One cannot expect the speed with which, say, a multiplication or a memory access, can be performed to increase indefinitely. So greater speed of computing

will come most strongly from being able to do many of these operations at once. Some argue that having a big machine that can do a problem fast is not economically advantageous over having several slower small machines. But isn't it much better to have one machine that takes a day for a problem than 100 which take 100 days for a problem?

Of course, this isn't the most important point in the speed question. The more important "intelligence" quality of a fancy machine with a good internal programming library will be mentioned later.

2. *Expansibility.* A machine based on parallel control can be designed so that it can be readily expanded, more or less indefinitely, by the addition of more arithmetic, memory, control, etc. units. This could be a very desirable feature especially when a valuable library of sub-routines, etc., had been built up.

3. *Self-maintenance.* A parallel control machine could locate points in itself needing repair. It would tend not to be completely incapacitated by any single material failure.

4. *Ease of programming, the "intelligence" quality, ability to handle complex problems efficiently.* We could define the intelligence of a machine in terms of the time needed to do a typical problem *and* the time needed for the programmer to instruct the machine to do it. Suppose a human can do a problem in 3 hours and the machine in 1 minute, but programming takes 5 hours. Then the machine has no advantage for the isolated problem, and would be worthwhile only when many similar problems are to be done. Simplifying the programming requirements so that the machines can take less explicit instructions is extremely important for the development of the art. This does not require special design of the machine to take instructions more readily but rather the development of interpretative programs by which translation from more abstract into more explicit instructions can be effected by the machine itself. This process is capable of indefinite extension. Its cost is in machine time, either spent in constructing an explicit program before computing or in translating during computation from abstract to explicit instructions. Here parallel control offers

a distinct advantage since it permits the translation to occur simultaneously with the computation, indeed several different stages of translation might be occurring simultaneously.

Before saying any more about the advantages of parallel control we should give a more concrete idea of a design for a parallel control computer. Consider a machine having

(a) logical units (control units)
(b) arithmetical units
(c) fast memory units
(d) slow memory units
(e) input units
(f) output units
(g) the communication network.

The more routine aspects of control are separated from the rest. Each arithmetic unit is, for example, to have the capacity to follow a sequence of orders stored in the fast memory so long as no decisions are required. A decision is made when the sequence of orders followed depends on the result of following them to some point where two or more possible continuations would exist, the continuation followed depending on that result. Division should not be a decision requiring process; the arithmetic unit should divide in response to a single order.

When a decision must be made the relevant information is referred to a logical unit. Also the logical units should handle all information processing that is not properly an arithmetical operation or an input or output operation. In this category would fall program translation, etc. The design of the machine should make it unnecessary for a logical unit to refer to an arithmetic unit in performing any purely logical task, such as manipulating orders, modifying programs, searching the memory.

A block diagram of such a machine would be as shown in the figure. For different applications, the balance between the numbers of units of various types would be different. It would generally be desirable to have fairly many fast memory units to minimize waiting in line for access.

JOHN F. NASH, JR.

Parallel Control

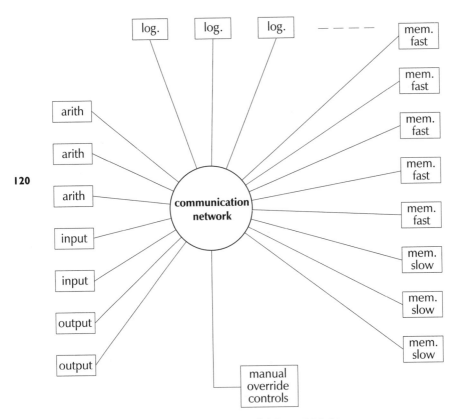

Block Diagram of a Parallel Control Machine

It is quite likely that for certain purposes, one would want specialization within the classification "logical unit" or the other types of units. For example, for a translating machine an arithmetical unit is pointless but specialized input readers would be valuable.

Now let us consider what functions the various types of units should perform.

1. *Arithmetic Unit.* Should have the capacity to obey commands to:
   (a) perform an arithmetic operation,
   (b) request a specified number from a specified memory unit and place it for operation,
   (c) send out the result of an arithmetic operation to a specified memory unit to be stored in a specified location,

(d) refer to a specified memory locus for a new command, and then continue in canonical sequence obeying the commands stored at following memory loci,

(e) be responsive (or unresponsive) to an overriding command transmitted from a logical unit,

(f) stop,

(g) "report" to specified logical unit.

2. *Fast Memory Unit.* The fast memory units are to be passive and respond to direct requests for specified stored numbers from other units or to direct instructions to store a transmitted number at a specified locus. So their functions are described as functions of the other units.

3. *Slow Memory Unit.* Obeys commands in the same manner as an arithmetic unit but has different functions, these are to read from a fast memory a specified number or a specified sequence of numbers in canonical order or to do the reverse, sending numbers into a fast memory.

4. *Input and Output Units.* Follow commands in the same manner as an arithmetic unit and act between fast memory units and forms of communication with the exterior of the machine.

5. *Communicating Network.* This apparatus is analogous to an automatic telephone system. It must transmit orders and data from one unit to another as directed by the unit sending the order or direction. It must have a provision for dealing with two or more messages directed simultaneously to the same unit so that only one is permitted at a time. If a unit is delayed in getting its message through, it waits until the message goes through unless it is acting under a command that instructs it to some alternative in such a case. This is a "decision," so this would apply only to a logical unit. All other types would wait.

6. *Logical Unit.* In this system, all powers not delegated to the other types of unit are reserved to the logical units. They may be considerably more richly equipped with powers to perform various specific logical operations than are the control units of current machines. A logical unit should have enough basic operations so as not to require the help of an arithmetic unit in

building up (efficiently) more complex or special logical functions or decisions. The writer cannot give anything close to an optimal or efficient set of basic operations now, but can illustrate a set that gives an idea of how the logical units might function.

Assume a logical unit can contain at one time seven numbers (which term will include data in the form of a set of binary digits regardless of whether or not it is to be interpreted as a number) stored at places a, b, c, α, β, γ, δ. The numbers α, β and δ are to be thought of as data. a, b, and c are commands. γ is partly like data and partly like a command, as we shall see. It can be thought of as a subsidiary command sometimes given meaning by c. The command c is the operating command determining the current operation of the unit.

In the schematic shown, dotted lines indicate influence or control. c and δ control the decision on which command a or b is to succeed c, if either. This is the basic decision operation. c and γ control the production of a number δ as a logical function of α and β. Think of c as commanding that a logical function of a certain type be computed from α and β and of γ as determining the specific function of that type.

We must assume here (really this is an assumption for simplicity) that numbers are longer than addresses so that a number can contain

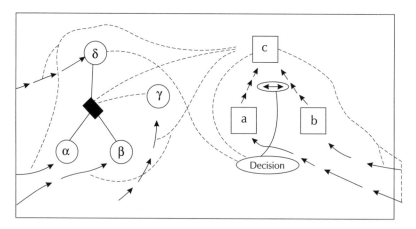

Schematic of Logical Unit Function

both instructions and a memory locus. Consider the following set of commands (c is the command):

(A) Replace c by the number in memory locus _____. (When no rule for replacing c is given explicitly in a command, c is replaced by the next command in canonical sequence.)

(B) Replace c by a if δ has property _____, otherwise by b. The property of having first digit 1 would be enough, but it might be more efficient to have some other decisions built in as well.

(C) Compute the logical function of type _____ described by γ of the numbers α and β and place the result as δ.

(D) Send δ to memory locus _____.

(E) Send δ to arithmetic unit _____ as an overriding command.

(F) Send δ to logical unit _____, etc. (e.g., to other unit types).

(G) Send δ to logical unit _____ as a datum. (As distinguished from a command.)

(H) Be responsive to a command received from another logical unit.

(I) Accept a command received from another logical unit for storage as a.

(J) Reject commands received.

H, I, J are commands that remain in effect until countermanded (by H, I, or J). When a number δ is sent to another unit and rejected by that unit, the sending unit must follow command b, otherwise the next command in sequence as usual.

(K) Be receptive to a datum sent from another logical unit and store it as β.

(L) Be unreceptive.

(M) Place the number in memory locus _____ in position as α.

(M') . . . as β.

(M'') . . . as γ.

(M''') . . . as a.

(M'''') . . . as b.

(N) Stop being responsive to commands or reports.

(O) Be responsive to report from an arithmetic or other lower unit. Such a report contains the address of a command and the logical unit obtains that command from memory and places it as c.

(P) Be unresponsive to reports.

Now we can consider what logical functions should be included under (c).

(c1) Produce δ as the number which has the same digits as α in the places where γ has 1's and the same as β where γ has zeros.

(c2) Produce δ as the number which has 1's where α and β have the same digit and zeros where they differ.

(c3) Produce δ by permuting the digits of α where γ has 1's one step cyclically.

(c4) Produce δ by having a zero where γ has a zero and where γ has a 1 by having a 1 unless the corresponding digits of both α and β are 1's (Non-Conjunction).

The attempt at a specific illustration above is, of course, nothing more than that. It's just to give a picture of the functions contemplated for logical and arithmetic and other units in a parallel control machine.

Of course, decision making does not have to be set up as a specific function in a logical unit. Decisions can be made by taking data numbers, transforming them into order numbers, and using them as orders. The programming could be set up to use this device, which would be slower.

The computation of logical functions could be separated from the other functions of a logical unit. It is not clear whether or not this would be advantageous.

Ultimate Advantages

Perhaps this type of machine organization would have the most value for very large machines intended to have a wide range of applicability. Consider a large machine which is to be built at a time (in the future) when the costs of computation have decreased, when very large fast memories are feasible, when printed circuits etc., can be mass produced automatically. When such a development exists, one will want the human labor by mathematicians and programmers needed in solving a problem reduced to minimum. The mathematician should have to do little more than state the problem and general method of computation to the machine in almost ordinary mathematical style. The machine should have considerable interpretative capacity and should itself work out the program of computation, descending level by level of abstraction to the level of specific orders and storage patterns for the computation.

To do this sort of interpretative work the machine must have a quite large general interpretation program stored within it in readiness for new problems. In principle current machines could do this, though not necessarily very efficiently or rapidly, if they had a sufficiently developed program for the purpose stored in their memories. But an interpretative program adequate to essentially eliminate programming would be too much for their memories. Now we can see how the large parallel control machine capable of handling several problems simultaneously would have an advantage. The interpretation program would only need be represented once. So a big advantage of one big machine is that it can afford to store much more basic information, and especially store more with rapid accessibility, than could any one of a collection of smaller machines.

If logical units could be made cheap enough, and the same for memory, one could afford to have trial and error and search processes used in the machine. Trial and error processes and learning processes (which would require a lot of memory) would be helpful in developing high order interpretative capacity. Ultimately trial and error processes, combined with search or association processes, abstraction processes, and learning or conditioning processes should lead to the learning machine or even to the genuine thinking machine. For this development parallel logical operation will certainly be important.

It is interesting to consider what a thinking machine will be like. It seems clear that as soon as the machines become able to solve intellectual problems of the highest difficulty which can be solved by humans they will be able to solve most of the problems enormously faster than a human.

In closing, the human brain is a highly parallel setup. It has to be.

JOHN F. NASH, JR.

Parallel Control

CHAPTER 10

J O H N   F .   N A S H ,   J R .

Real Algebraic Manifolds

(Received October 8, 1951)

Introduction

The main purpose of this paper is to develop some connections between differential geometry and real algebraic geometry. These connections concern the geometrical forms that can be assumed by a real algebraic variety.

Our first theorem asserts that any closed differentiable manifold can be displayed as a non-singular portion of a real algebraic variety. More specifically, any differentiable imbedding of the manifold may be approximated by such an algebraic imbedding of it.

The existence of these algebraic models of differentiable manifolds suggested the formulation of an abstract concept of a differentiable manifold with an algebraic structure. Part of the purpose of this concept is to reveal the relationship between the various possible algebraic representations of a manifold. For it turns out that there is essentially only one algebraic structure (of the type we consider) that can be put on a given closed differentiable manifold.

We place an algebraic structure on a differentiable manifold by distinguishing certain functions on it as algebraic functions. The class of algebraic functions is required to satisfy a few reasonable conditions. The choice of the class of functions which is to be regarded as the class of algebraic functions is not uniquely determined; but if any two such classes are given, each of which satisfies the conditions, then either may be transformed into the other by a differentiable homeomorphism of the manifold into itself. It is in this sense that the algebraic structure is uniquely determined. The combination of the manifold and the class of algebraic functions is called a real algebraic manifold.

It is convenient to use the notion of a basic set of functions on a closed analytic manifold in stating the precise definition of a real algebraic manifold. A *basic set* is a finite set of single valued real functions analytic over the entire manifold; they are required to be such that they could be used as imbedding functions to imbed the manifold in a Euclidean space (non-singularly). [To make clear what is meant by the term 'imbedding functions': Let $\mathfrak{M}$ be the manifold, p the symbol for a point of $\mathfrak{M}$, and $x_1(p)$, $x_2(p)$, $\cdots$, $x_n(p)$ the Cartesian coordinates of the image of p under the imbedding of $\mathfrak{M}$ into E^n. The x_i's are the imbedding functions.]

We define a *real algebraic manifold* as a closed analytic manifold $\mathfrak{M}$ plus a ring $\mathfrak{R}$ of functions on it such that:

(a) Each function of $\mathfrak{R}$ is a single valued real function analytic at all points of $\mathfrak{M}$.

(b) There is a basic set on $\mathfrak{M}$ composed of functions in $\mathfrak{R}$.

(c) If a set of functions in $\mathfrak{R}$ contains more functions than the number of dimensions of $\mathfrak{M}$ then the functions of this set must satisfy some non-trivial polynomial dependence relation. Thus the transcendence degree of $\mathfrak{R}$ must be the dimensionality of $\mathfrak{M}$.

(d) Finally, $\mathfrak{R}$ must be maximal within the class of rings satisfying the above conditions.

This concept of a real algebraic manifold is to a certain extent similar to the idea of a complex algebraic manifold. In the complex case one would use single valued meromorphic functions and a basic set would define a mapping of the manifold into a complex projective space. Here the class of functions would form a field since the reciprocal of a meromorphic function (except 0) is also meromorphic.

The ring associated with a real algebraic manifold (regarded as an abstract algebraic object) is actually sufficient to determine the manifold itself. One can construct the manifold by taking the maximal ideals of the ring to be its points. And conversely, given a closed differentiable manifold there is only one possible abstract ring that can be associated with it. These properties are the compensations of our rings for not being fields.

Our definition of a real algebraic manifold specifically states that the manifold $\mathfrak{M}$ can be imbedded in some Euclidean space using imbedding functions that are in its ring $\mathfrak{R}$ of algebraic analytic functions. Such an imbedding will be called a *representation* of the algebraic manifold $(\mathfrak{M}, \mathfrak{R})$.

We allow ourselves some flexibility in the use of the term "representation." It will be used in two senses which, from a logical viewpoint, are not equivalent. In the first sense a representation is a mapping of an algebraic manifold into a Euclidean space. In the second sense, a representation is a point set in a Euclidean space that is the image of such a mapping. The use of the term in the second sense is a sort of logical short-cut. Furthermore, if an imbedding of a differentiable or analytic manifold happens to be a representation in this second sense we shall call it an algebraic imbedding, or algebraic representation of the original manifold.

The real algebraic manifolds we have defined should not be confused with algebraic varieties. It is their representations that are closely related to varieties. A representation of an n-dimensional algebraic manifold will be a portion of an n-dimensional variety. Part of our work is devoted to characterizing the portions of real varieties which are representations.

To make this characterization we introduce the geometrical-analytical concept of a sheet of a real variety. A *sheet* is a subset of a real variety and must have the properties:

(a) Given any two points in the sheet, there is an analytic path which joins them and lies entirely within the sheet. By an analytic path we mean the image of the unit interval, $t = 0$ to $t = 1$, under an analytic mapping into the space of the variety. A path can have self-intersections and cusps, but each coordinate must be an analytic function of the time, t.

(b) The sheet must be maximal in the class of subsets of the variety which have property (a).

(c) There must exist a point of the sheet which has a neighborhood in which all points of the variety are also points of the sheet.

In this paper we actually work with sheets that are also analytic manifolds imbedded in the space. But there are some interesting conjectures that can be made about the sheets with singularities. We mention some of these later.

The compact non-singular sheets, that is, those which are closed analytic manifolds, are representations of algebraic manifolds. And conversely, an algebraic representation of a manifold is always a sheet of a variety. This is our characterization of representations.

If a sheet is topologically separate from the other parts of the variety we call it an *isolated sheet*. To be precise, an isolated sheet is one which can be surrounded by an open set containing no other points of the variety.

Now an algebraic representation of a manifold is more satisfying if it is an isolated sheet of a variety. We call this more elegant representation a *proper representation*. Given a closed differentiable n-manifold, one can always obtain a proper representation of it in E^{2n+1}. Unfortunately, we have not succeeded in proving an approximation theorem for proper representations.

A paper [2] by H. Seifert contains some results on algebraic representations of manifolds. He considers manifolds which are differentiably inbedded in a Euclidean space in such a way that the bundle of normal vectors is a product bundle. This is simply the necessary and sufficient condition for there to be a set of $n - r$ functions of class C^1 which vanish on the manifold and which have gradients that are orthogonal unit normal vectors at each point of the manifold, where n is the dimensionality of the Euclidean space and r that of the manifold. A set of $n - r$ sufficiently good polynomial approximations to these functions will define an approximating proper algebraic representation of the manifold. The algebraic approximation is part of the locus where all $n - r$ approximating polynomials vanish simultaneously. The approximation of the original functions by the polynomials must be of the second order, that is, the first derivatives of the polynomials must approximate the corresponding first derivatives of the original functions. Seifert's method has the advantage of obtaining a proper representation but it can be applied only to a certain portion of the class of orientable manifolds.

We think there are good opportunities for further work on the algebraic representation of manifolds. But it seems that a new approach will be needed, for we have tried to extend our results in several different directions without success. Some of the natural conjectures are given at the end of this paper.

There is one more concept we must define here. This is the concept of equivalence of two real algebraic manifolds. An *algebraic correspondence* between two algebraic manifolds $(\mathfrak{M}_1, \mathfrak{R}_1)$ and $(\mathfrak{M}_2, \mathfrak{R}_2)$ is an isomorphism ϕ between $\mathfrak{R}_1$ and $\mathfrak{R}_2$. When such a correspondence exists between two algebraic manifolds, they are called *equivalent*. This definition is stronger than it seems; we shall show that such an algebraic correspondence always induces an analytic homeomorphism between the manifolds.

LEMMA I. *If $\mathfrak{M}$ is a manifold analytically imbedded in E^n it has a neighborhood $\mathfrak{N}$ in E^n in which there is a unique nearest point y on $\mathfrak{M}$ for each point x of $\mathfrak{N}$, and in which y depends analytically on x.*

Proof. Let $\lambda_1, \lambda_2, \cdots \lambda_r$ parametrize analytically and without singularities a neighborhood N_1 in $\mathfrak{M}$. Let $y_i(\lambda)$ be the imbedding functions. Thus the matrix

$$\left\| \frac{\partial y_i}{\partial \lambda_\alpha} \right\|$$

has rank r at all points of N_1.

Consider the system

$$x_i = z_i + y_i(\lambda)$$
$$\mu_\alpha = \sum_i z_i \frac{\partial y_i}{\partial \lambda_\alpha}.$$

Here x is to be thought of as a point in E^n and z as the vector from $y(\lambda)$, which is on $\mathfrak{M}$, to x. If $\mu = 0$ then z is normal to $\mathfrak{M}$.

We now investigate the transformation $T : (z, \lambda) \to (x, \mu)$. The Jacobian of T is the determinant

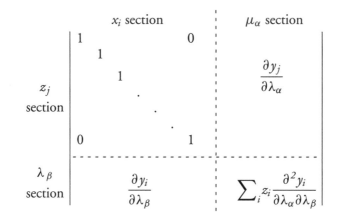

If $z = 0$ it is easy to see that the absolute value of the Jacobian is the sum of the squares of the r by r minors of the matrix

$$\left\| \frac{\partial y_i}{\partial \lambda_\alpha} \right\|,$$

and therefore non-zero. So in the vicinity of a point $(0, \lambda^{(0)})$ in the (z, λ) space, the transformation T is non-singular.

Such a point $(0, \lambda^{(0)})$ corresponds to a point $(x^{(0)}, 0)$ in the (x, μ) space, and $x^{(0)}$ is on $\mathfrak{M}$ and in N_1. The point $(x^{(0)}, 0)$ must have a neighborhood N_2 in the (x, μ) space which is the one-one image of a neighborhood of $(0, \lambda^{(0)})$ in which T is non-singular.

Using the implicit function theorem, we see that in N_2 analytic inverse functions $z(x, \mu)$ and $\lambda(x, \mu)$ will exist. For each point x such that $(x, 0)$ is in N_2 we may define a point $y(x)$ on $\mathfrak{M}$ by $y(x) = y(\lambda(x, 0))$. The point $y(x)$ is such that the line segment from $y(x)$ to x is normal to $\mathfrak{M}$. Also $y(x^{(0)})$ is $x^{(0)}$. Let N_3 be the neighborhood of $x^{(0)}$ in which $(x, 0)$ is in N_2 and in which $y(x)$ is defined.

We shall now show that if x is near enough to $x^{(0)}$ then $y(x)$ is the unique nearest point of x on $\mathfrak{M}$. First, observe that if x is near enough to $x^{(0)}$ all nearest points on $\mathfrak{M}$ to x must be in N_1. Second, note that if λ is the parametrization of a nearest point, then z, which is the vector from $y(\lambda)$ to x, is normal to $\mathfrak{M}$ so that $T(z, \lambda)$ is $(x, 0)$.

But (z, λ) must be near $(0, \lambda^{(0)})$ if x is near $x^{(0)}$ and the transformation T is locally one-one at $(0, \lambda^{(0)})$. Thus sufficiently near $x^{(0)}$ any point of E^n will have a unique nearest point on $\mathfrak{M}$ which will depend on it analytically.

Since $x^{(0)}$ could have been any point on $\mathfrak{M}$ the assertion is evidently true.

LEMMA 2. *Let f be a real single valued function analytic in an open set $\mathfrak{E}$ of E^n, where $\mathfrak{E}$ contains the closed region $\mathfrak{A}$ bounded by a closed analytic hypermanifold $\mathfrak{B}$ ($\mathfrak{B}$ may be disconnected.). Then there is a sequence $\{f_v\}$ of polynomials in the coordinates of E^n which converge uniformly to f in $\mathfrak{A}$. Moreover, for any type of derivative $\mathbf{D}$ with respect to the coordinates of E^n, such as,*

$$\mathbf{D} = \frac{\partial^{17}}{\partial x_3^4 \, \partial x_5 \, \partial x_2^{11} \, \partial x_{13}},$$

$\{\mathbf{D} f_v\}$ converges to $\mathbf{D} f$ uniformly in $\mathfrak{A}$.

Proof. Let $p(x)$ be the distance of a point x from the region $\mathfrak{A}$. Given any positive constant δ we may define a continuous function $\zeta(x)$ by the equations

(a) $\zeta = 0$ if $\rho \geqq \delta$,

(b) $\zeta = 1$ if $\rho = 0$,   and

(c) $\zeta = \dfrac{e^{-1/(\delta - \rho)}}{e^{-1/(\delta - \rho)} + e^{-1/\rho}}$ if $0 < \rho < \delta$.

Let us assume δ is small enough to make ζ a C^∞ function and to make $\zeta = 0$ at all points outside of $\mathfrak{E}$. This can be since Lemma 1 shows that ρ is analytic near $\mathfrak{A}$.

Consider the function ζf, and think of its range of definition as extended to all E^n by making it zero when f is undefined. In a cube large enough to contain all points where $\zeta f \neq 0$ one may obtain a very well behaved multiple Fourier series for ζf. The coefficients will decrease faster than any negative power of the order of the terms to which they correspond. This will make the partial sums, $\{S_n\}$, converge uniformly to the function, and for each type of derivative $\mathbf{D}$ the sequence $\{\mathbf{D} S_n\}$ will converge uniformly to $\mathbf{D}(\zeta f)$.

Each partial sum S_n is an entire function of the several (for this purpose complex) variables and therefore has a very rapidly convergent power series. For any S_n the partial sums of its power series will converge uniformly in $\mathfrak{A}$ to S_n and their derivatives will converge uniformly to the derivatives of S_n.

It now becomes clear that by using a diagonal process we can get a sequence of polynomials which converge to f in $\mathfrak{A}$ in the manner specified in the theorem, since $\zeta f = f$ in $\mathfrak{A}$.

THEOREM 1. *A differentiable imbedding $\mathfrak{D}$ in E^n of a closed differentiable manifold may be approximated by an algebraic representation of the manifold.*

Our construction of the algebraic approximation is similar, in essence, to the method Whitney [1] used in obtaining in analytic approximation to a differentiable manifold inbedded in a Euclidean space.

Let us give an example which shows the main features of our construction process. Suppose $\mathfrak{D}$ is a 5-manifold in E^8. The manifold $\mathfrak{D}$ can be assumed to be an analytic imbedding since the problem can be reduced to this case. We work in a small surrounding neighborhood $\mathfrak{N}$ of $\mathfrak{D}$ where each point x of E^8 has a unique nearest point $y(x)$ on $\mathfrak{D}$. We obtain a vector function $u(x)$ whose components are polynomials and which approximates the function $y(x) - x$. We use algebraic functions of the coordinates of E^8 to describe a 3-plane $(3 = 8 - 5)\mathfrak{P}(x)$ passing through x which is approximately the 3-plane $\mathfrak{R}(x)$ that is normal to $\mathfrak{D}$ at $y(x)$.

The condition which defines the algebraically imbedded approximating manifold is that u should be normal to $\mathfrak{P}$. This amounts to 3 algebraic conditions, since $\dim(\mathfrak{P}) = 3$, so it should hold on a 5-manifold in E^8. For this to be a good algebraic approximation to $\mathfrak{D}$, however, u must not only approximate $y(x) - x$ but should also have derivatives approximating those of $y(x) - x$.

Conventions

Before actually giving the proof of Theorem 1 we must explain some terminological conventions that are used to shorten and simplify it.

Certain letters $x, y, z, u, v, \phi, \varphi$, and ψ will stand for vectors or positions. When one of these carries a subscript this indicates one of its components. The vector x represents position in E^n; all derivatives of functions which we shall use will be taken with respect to the Cartesian coordinates of E^n, that is the components of x. The letters **M, L, P, Π,** and **K** will denote matrices (operators).

Another set of conventions serves to implement a conceptual device that enables us to carry out the discussion of approximation questions without using a horde of inequalities. These conventions and the conceptual device are best explained by using in part the method of illustration. The explanation follows below.

In the following proof all functions of x will be considered only in a certain region of E^n which will be called $\mathfrak{N}$. Now suppose f is a function analytic in $\mathfrak{N}$ and that we want to approximate it by a polynomial p. It turns out to be more convenient to talk about a sequence $\{p_\nu\}$ of improving polynomial approximations to f. For notational symmetry we introduce a sequence $\{f_\nu\}$ of functions, all equal to f.

Now the basic idea is to study the approximation relationship between sequences of functions rather than between single functions. This leads to a concept that is more easily handled with precision. For example we write

$$\{f_\nu\} \approx \{p_\nu\}$$

to say that the indicated function sequences approximate each other. And this means that (a) $\{f_\nu - p_\nu\}$ must be a sequence of functions which converge uniformly to zero in the domain $\mathfrak{N}$ of discussion; and (b) for any type of derivative **D** with respect to the Cartesian coordinates of E_n the sequence $\{\mathbf{D}f_\nu - \mathbf{D}p_\nu\}$ must converge uniformly to zero in $\mathfrak{N}$.

Thus our approximation concept involves approximation of derivatives too. This definition has the advantage that differentiation of the functions in two function sequences which approximate each other yields two new mutually approximating sequences.

Since we shall use this convention throughout the proof it is very convenient to make it a convention that a symbol, such as f, representing a function will always stand for a sequence of functions, such as $\{f_\nu\}$. Sometimes all members of the sequence will be the same, sometimes not. We shall also employ the obvious generalizations of this technique to vector functions, polynomials, operators, etc. The general symbol for the approximation relationship will be $\approx$.

On occasion, we shall use $\approx$ to indicate approximation between objects which do not of themselves have derivatives, such as unordered sets of numbers. When used in this way the necessary modification of the meaning of $\approx$ which must be understood should be clear from the context.

In the proof all statements will be made as if for a particular value of the general index ν which indexes each function sequence, and thus as if each symbol stood for a single function instead of a sequence of them. And we shall use the convention that statements and definitions are required to be valid or meaningful only for almost all values of ν, i.e., only when the approximations are sufficiently good. For example, a definition may be meaningless unless certain quantities are near enough to zero. How this convention operates in detail will become clearer as the proof proceeds.

Proof of Theorem 1. The first step in the proof is to use Whitney's analytic approximation result [1] and assume $\mathfrak{D}$ is analytically imbedded, since otherwise it could be approximated by a differentiably homeomorphic analytic manifold, and an algebraic approximation to the new manifold would approximate the original one.

Our first lemma insures that $\mathfrak{D}$ will have a neighborhood $\mathfrak{N}$, which we take for convenience to be a closed spherical surrounding of $\mathfrak{D}$, such that in $\mathfrak{N}$ each point x has a unique nearest point $y(x)$ on $\mathfrak{D}$. Moreover $y(x)$ is to be an analytic function in $\mathfrak{N}$.

Let v stand for the vector function $y - x$. Let r be the dimensionality of $\mathfrak{D}$. There will be a unique $(n-r)$-plane $\mathfrak{K}$ perpendicular to $\mathfrak{D}$ at y. This varies analytically with y, and hence with x. Let $\mathbf{K}$ be the projection operator which projects all vectors into their components parallel to $\mathfrak{K}$. For example $\mathbf{K}v = v$. $\mathbf{K}$ and v are now analytic functions of x in $\mathfrak{N}$.

The next step is to obtain polynomial approximations to v and $\mathbf{K}$ (actually we obtain a sequence of improving approximations for each). Our second lemma was proved to enable us to do this. So let u be the approximation to v and $\mathbf{L}$ the approximation to $\mathbf{K}$ (u and $\mathbf{L}$ really represent sequences of functions). In symbols, $u \approx v$ and $\mathbf{L} \approx \mathbf{K}$. It is convenient for **L** to be a symmetric matrix, a condition which can obviously be fulfilled since **L** is to approximate the symmetric matrix **K**. So we shall assume **L** is symmetric.

Let $a(\lambda)$ and $\alpha(\lambda)$ be the characteristic polynomials of **K** and **L**, respectively, normalized to have first coefficient $+1$. We have $a(\lambda) \approx \alpha(\lambda)$ and therefore

$$[\text{the set of roots of } a] \approx [\text{the set of roots of } \alpha].$$

Now $a(\lambda) = \lambda^r (\lambda - 1)^{n-r}$, since **K** projects vectors onto the $(n - r)$-plane $\Re$. Consequently the set of roots of either a or α may be divided into two parts; first a set of r roots ≈ 0, and second a set of $n - r$ roots ≈ 1. Let b and β be the factors of a and α which embrace the small roots and let c and γ be the complementary factors (all with first coefficient $+1$).

The coefficients of β will be analytic functions of x. We pause now to prove this via a little lemma which tells us when the coefficients of a factor of a polynomial will depend analytically on those of the original polynomial.

From r real numbers $\delta_1, \delta_2, \cdots, \delta_r$ and $n - r$ real numbers $\theta_1, \theta_2, \cdots, \theta_{n-r}$ we may form two polynomials,

$$\Delta = \lambda^r + \delta_1 \lambda^{r-1} + \cdots + \delta_r$$

and

$$\Theta = \lambda^{n-r} + \theta_1 \lambda^{n-r-1} + \cdots + \theta_{n-r}.$$

Their product $\Delta\Theta$ may be written

$$\Delta\Theta = \lambda^n + \xi_1 \lambda^{n-1} + \cdots + \xi_n.$$

The transformation taking the δ's and θ's into the ξ's is clearly an analytic one, for the ξ's are polynomials in the δ's and θ's. We wish to find the conditions under which there is locally an analytic inverse transformation. The obvious approach is to look at the Jacobian of the transformation,

$$\frac{\partial(\xi_1, \cdots, \xi_n)}{\partial(\delta_1, \cdots \delta_r, \theta_1, \cdots \theta_{n-r})}.$$

This Jacobian turns out to be just the resultant of the polynomials Δ and Θ. Since the resultant of two polynomials vanishes only when they have a common factor the transformation will be non-singular and there will be a local analytic inverse transformation unless Δ and Θ have a common root. Thus the δ's and θ's will depend analytically on the ξ's in an inverse relationship except when Δ and Θ have a common root.

Now in the case of β and γ the roots of β approximate 0 while those of γ approximate 1. Thus β and γ have no root in common. And since their product is α, the coefficients of β depend analytically on those of α, and hence analytically on x. Actually the coefficients of β are also algebraic functions of x, that is, of the coordinates of E^n. Later we shall exhibit specific polynomials describing this dependence.

We may now define the operator $\mathbf{P} = \beta(\mathbf{L})$. We are using here the standard device of putting a matrix in a polynomial to obtain a new matrix. The fact that β is a factor of $\mathbf{L}$'s characteristic polynomial α makes $\mathbf{P}$ have special properties. Any eigenvector of $\mathbf{L}$ must also be an eigenvector of $\mathbf{P}$, but the associated eigenvalue will be altered. The new eigenvalue is the result obtained by putting the old one into the polynomial β. Since the r small eigenvalues of $\mathbf{L}$ are the roots of β they become r zero eigenvalues of $\mathbf{P}$. Since $\beta \approx b = \lambda^r$, the $n - r$ large eigenvalues of $\mathbf{L}$ which are approximately $+1$ will remain approximately $+1$. Also $\mathbf{P} \approx \mathbf{K}$ because $\mathbf{P} = \beta(\mathbf{L}) \approx b(\mathbf{K}) = \mathbf{K}^r = \mathbf{K}$.

The operator $\mathbf{K}$ may be thought of as projecting vectors into the $(n - r)$-plane $\mathfrak{R}$ which passes through x and intersects $\mathfrak{D}$ perpendicularly at $y(x)$. Similarly $\mathbf{P}$ defines an $(n-r)$-plane $\mathfrak{P}$ through x, since the rank of $\mathbf{P}$ is always $n - r$. However, the $n - r$ non-zero eigenvalues of $\mathbf{P}$ are only approximately and not exactly equal to 1 so that the action of $\mathbf{P}$ is not a pure orthogonal projection on $\mathfrak{P}$. There will be some distortion. Since $\mathbf{P} \approx \mathbf{K}$, we may write $\mathfrak{P} \approx \mathfrak{R}$ to indicate that $\mathfrak{P}$ is nearly parallel to $\mathfrak{R}$.

Now let

$$\phi = \mathbf{P}u$$

and

$$\psi = \mathbf{K}\phi.$$

$\phi \approx v$ because $\mathbf{P} \approx \mathbf{K}$ and $u \approx v$ so that $\mathbf{P}u \approx \mathbf{K}v = v$. And $\psi \approx v$ since $\mathbf{K}\phi \approx \mathbf{K}v = v$. Also $\phi = 0$ and $\psi = 0$ are really equivalent conditions, because ϕ lies in the plane $\mathfrak{P}$ which is nearly parallel to $\mathfrak{R}$ so that ϕ itself is almost in $\mathfrak{R}$. Thus $\mathbf{K}\phi$, which is ψ, will not vanish unless ϕ vanishes. $\phi = 0$ will be the condition that determines the algebraic approximation to $\mathfrak{D}$. The advantage in considering ψ is that it lies in the plane $\mathfrak{R}$ which is normal to $\mathfrak{D}$ at $y(x)$.

We define a vector function $z(x)$ by

$$z = x + v - \psi$$

z is to be thought of as a point in E^n. Because $v \approx \psi$ we have **139** $z \approx x$. This approximation relationship between z and x requires the corresponding derivatives of z and x to approximate each other. So, in particular, the Jacobian of the transformation $x \rightarrow z(x)$ will approximate that of the identity transformation $x \rightarrow x$, which has Jacobian $+1$. Consequently every point of $\mathfrak{N}$ will have a neighborhood in which the mapping $x \rightarrow z(x)$ is one-to-one and non-singular.

We must show that the mapping $x \rightarrow z(x)$ is one-to-one in the large, that is, over the whole region $\mathfrak{N}$. In proving this the fact that the boundary, which we shall call $\mathfrak{O}$, of $\mathfrak{N}$ is an analytic manifold is significant. (In this connection the usual requirement that a manifold be connected is relaxed.) $\mathfrak{N}$ is a *spherical* surrounding of $\mathfrak{D}$, so that $\mathfrak{O}$ is the set of points at a certain distance from $\mathfrak{D}$. $\mathfrak{O}$ is analytic, since the distance to $\mathfrak{D}$ (which is the function $|v|$) is a function that is analytic in $\mathfrak{N}$ except on $\mathfrak{D}$ itself.

Now consider the mapping $x \rightarrow z(x)$ restricted to $\mathfrak{O}$. Since this mapping approximates the identity mapping and has derivatives approximating those of the identity mapping, it is clear that it maps $\mathfrak{O}$ one-to-one into another analytic manifold $\mathfrak{O}^*$ which approximates $\mathfrak{O}$. It is now easy to see that because the mapping $x \rightarrow z(x)$ is locally one-to-one in $\mathfrak{N}$ and is one-to-one on the boundary of $\mathfrak{N}$ that it must be one-to-one over $\mathfrak{N}$ as a whole. Hence $\mathfrak{N}$ is mapped analytically into a similar region $\mathfrak{N}^*$ in which an inverse mapping $z \rightarrow x(z)$ is defined. (Of course all this may be true only for high enough values of the general index v of the function sequences which such symbols as $z(x)$ really represent. We have been using our convention which justifies making statements true only for almost all values of v quite strongly in the foregoing arguments.)

$\mathfrak{D}$ will be inside $\mathfrak{N}^*$ because the boundary of $\mathfrak{N}^*$ approximates the boundary of $\mathfrak{N}$ and only a fringe of points near the boundary of $\mathfrak{N}$ will be exceptions to the rule that points of $\mathfrak{N}$ will also be in $\mathfrak{N}^*$. Therefore the inverse function $x(z)$ will be defined on $\mathfrak{D}$. It maps $\mathfrak{D}$ into an analytic manifold we shall call $\mathfrak{B}$. It is clear that $\mathfrak{B}$ approximates $\mathfrak{D}$.

For a point x on $\mathfrak{B}$, $z(x)$ is $x + v - \psi$ and is on $\mathfrak{D}$. The sum $x + v$ is y and is the nearby intersection of $\mathfrak{D}$ with the plane $\mathfrak{K}$ which passes through x and is normal to $\mathfrak{D}$ at y. The point y will not be the only intersection of $\mathfrak{K}$ with $\mathfrak{D}$ but it will be the only one that is near x. Because ψ is a vector parallel to $\mathfrak{K}$ the point $z(x)$, which is $y + \psi$, must be in $\mathfrak{K}$. But since $z(x)$ must also be on $\mathfrak{D}$ we conclude that ψ is zero. Otherwise $z(x)$ could not both be near x and be on $\mathfrak{D}$. Thus $\psi = 0$ at all points of $\mathfrak{B}$.

Conversely, whenever $\psi(x)$ is zero, $z(x)$ is $x + v$, which is y and is on $\mathfrak{D}$ so that x is on $\mathfrak{B}$. Thus $\mathfrak{B}$ is characterized by the condition $\psi = 0$, or equivalently, $\phi = 0$.

$\mathfrak{B}$ is our algebraic approximation to $\mathfrak{D}$. But we still must show that it satisfies the formal definition of an algebraic representation. This will be obtained as an incidental conclusion from the following construction of a proper representation $\mathfrak{B}^*$ in E^{n+r} of the original manifold, where $\mathfrak{B}^*$ will project into $\mathfrak{B}$ in E^n.

Let us parametrize E^{n+r} by $(x, \mathfrak{b})$, where x is a point of E^n and $\mathfrak{b}$ stands for the last r coefficients $(\mathfrak{b}_1, \mathfrak{b}_2, \cdots \mathfrak{b}_r)$ of the polynomial

$$\mathfrak{b}(\lambda) = \lambda^r + \mathfrak{b}_1 \lambda^{r-1} + \cdots + \mathfrak{b}_r.$$

Thus a point in E^{n+r} is really just the combination of a point in E^n and a polynomial (for our purposes).

Let η be the vector formed by the r coefficients of the remainder from the division of $\mathfrak{b}(\lambda)$ into $\alpha(\lambda)$, i.e.

$$\eta(\lambda) = \eta_1 \lambda^{r-1} + \eta_2 \lambda^{r-2} + \cdots + \eta_r = \text{ remainder } [\alpha(\lambda)/\mathfrak{b}(\lambda)].$$

A vector η is associated with every point of E^{n+r}, in fact the η_i's are polynomials in the coordinates of E^{n+r} since the coefficients of $\alpha(\lambda)$ are polynomials in the coordinates of E^n (polynomials in x).

Let

$$\Pi = \mathfrak{b}(\mathbf{L}) \qquad \text{and let} \qquad \varphi = \Pi u.$$

It is important to observe that the components of φ are polynomials in x, $\mathfrak{b}$. We shall show that the system

$$\varphi = 0$$
$$\eta = 0$$

⊞

defines a variety in E^{n+r} which has an isolated sheet $\mathfrak{B}^*$ that is the desired proper algebraic representation of the original differentiable manifold.

For each point x of $\mathfrak{N}$, the polynomial $\beta(\lambda)$ is defined. (x, β) may be regarded as a point in E^{n+r}, and the mapping $x \to (x, \beta)$ is analytic. Restricted to $\mathfrak{B}$, this mapping takes $\mathfrak{B}$ into a homeomorphic analytic manifold $\mathfrak{B}^*$ in E^{n+r}.

The system (⊞) will hold on $\mathfrak{B}^*$ because on $\mathfrak{B}^*$ the polynomials $\mathfrak{b}$ and β are the same, so that φ is ϕ, and will vanish, while η must vanish since $\eta = 0$ is merely a way of saying that $\mathfrak{b}$ divides α, a property $\mathfrak{b}$ will possess on $\mathfrak{B}^*$ where it is equal to β, which was originally defined as a certain factor of α.

All that we now need to show is that in a sufficiently restricted surrounding neighborhood of $\mathfrak{B}^*$ the *only* points where the system (⊞) holds will be those on $\mathfrak{B}^*$.

$\beta(\lambda) \approx \mathfrak{b}(\lambda)$ and $\mathfrak{b}(\lambda)$ is λ^r, and $\mathfrak{b}(\lambda)$ equals $\beta(\lambda)$ on $\mathfrak{B}^*$; thus for points of E^{n+r} near $\mathfrak{B}^*$ all coefficients of $\mathfrak{b}$ will be near zero, except the first coefficient, which is $+1$. Consequently all the roots of $\mathfrak{b}$ will be near to zero. Now the polynomial $\alpha(\lambda)$ has r roots near zero and $n - r$ roots near one. So if $\mathfrak{b}$ divides α, which is what is really meant by $\eta = 0$, and the point of E^{n+r} under consideration is near $\mathfrak{B}^*$ we know that $\mathfrak{b}$ must be the factor β of α that embraces the r small roots of α.

Thus we see that for points of E^{n+r} that are near $\mathfrak{B}^*$ and for which $\eta = 0$ the polynomial $\mathfrak{b}$ will be β; and from this it follows that at such points $\varphi = \phi$. If in addition $\varphi = 0$ then our point is of the form (x, β) and such that $\phi(x) = 0$; obviously such a point is on $\mathfrak{B}^*$. This completes the proof that $\mathfrak{B}^*$ is an isolated portion, which is clearly a sheet, of the variety defined by the system (⊞).

By Theorem 6, $\mathfrak{B}^*$ is a proper representation of the original manifold. Since the mapping $(x, \mathfrak{b}) \to x$ of E^{n+r} into E^n takes $\mathfrak{B}^*$ into $\mathfrak{B}$ it is clear that $\mathfrak{B}$ will also satisfy the requirements given in the definition of an algebraic representation. We can use Theorem 5 to show that $\mathfrak{B}^*$ will be a sheet of an irreducible r-dimensional variety in E^{n+r}. Consequently, one can obtain a proper representation in E^{2r+1} by the classical algebraic geometrical method of generic linear projection. Such a projection introduces no new singularities (i.e., no new double points) in the variety if the variety is in a space of dimensionality more than one plus twice its own dimensionality. We state the theorem we have just proved below.

THEOREM 2. *A closed differentiable manifold always has a proper algebraic representation in the Euclidean space of one more than twice its number of dimensions.*

THEOREM 3. *In an algebraic manifold $(\mathfrak{M}, \mathfrak{R})$, the points of $\mathfrak{M}$ are in natural biunique correspondence with the maximal ideals of $\mathfrak{R}$, a point corresponding with the ideal of all functions that vanish there.*

Proof. First, for any proper ideal $\mathfrak{g}$ of $\mathfrak{R}$, there is some point of $\mathfrak{M}$ at which all functions in $\mathfrak{g}$ vanish. For suppose otherwise: then each point is in the open set where some function of $\mathfrak{g}$ does not vanish. Since the class of all such open sets forms a covering of the compact space $\mathfrak{M}$, some finite subclass must also cover $\mathfrak{M}$. And if we square and add the functions matching the sets of such a finite subcovering, we obtain a function in $\mathfrak{g}$ that vanishes nowhere. But since the ideal $\mathfrak{g}$ cannot contain such a function (whose reciprocal must be in $\mathfrak{R}$) without containing all functions of $\mathfrak{R}$ we have established the initial assertion.

Second, for any point p of $\mathfrak{M}$ let $\mathfrak{g}_p$ be the ideal of all functions of $\mathfrak{R}$ that vanish at p; then p is the only point where all functions of $\mathfrak{g}_p$ vanish. Because remember that if p' is any other point there is a function f in $\mathfrak{R}$ which takes on different values at p and p'. And since all constant functions are in $\mathfrak{R}$, f minus the value of f at p is a function in $\mathfrak{g}_p$ that is not zero at p'.

By combining our first and second assertions one obtains the theorem quite easily.

THEOREM 4. *An algebraic correspondence ϕ between two algebraic manifolds $(\mathfrak{M}_1, \mathfrak{R}_1)$ and $(\mathfrak{M}_2, \mathfrak{R}_2)$ induces naturally an analytic homeomorphism between $\mathfrak{M}_1$ and $\mathfrak{M}_2$.*

Proof. Let S_1 and S_2 be basic sets on $\mathfrak{M}_1$ and $\mathfrak{M}_2$ composed of functions taken from $\mathfrak{R}_1$ and $\mathfrak{R}_2$. Let ϕS_1 be the set of functions of $\mathfrak{R}_2$ corresponding to the set S_1 in R_1, and likewise for ϕS_2. Then $T_1 = S_1 \cup \phi S_2$ and $T_2 = S_2 \cup \phi S_1$ will be corresponding basic sets on $\mathfrak{M}_1$ and $\mathfrak{M}_2$.

By Theorem 3, ϕ induces a correspondence of points of $\mathfrak{M}_1$ and $\mathfrak{M}_2$. Now the value of a function of $\mathfrak{R}_1$ (or $\mathfrak{R}_2$) at a maximal ideal (point) of functions is expressed in the algebraic structure of the ring. Hence the values of the functions of T_1 at a point in $\mathfrak{M}_1$ are the same as those of the functions of T_2 at the corresponding point in $\mathfrak{M}_2$.

Finally, since a basic set defines an analytic mapping of a manifold into a Euclidean space, T_1 and T_2 will send corresponding points of $\mathfrak{M}_1$ and $\mathfrak{M}_2$ into the same point of the common image, which is an analytically imbedded manifold. Hence the correspondence of points of $\mathfrak{M}_1$ and $\mathfrak{M}_2$ that ϕ induces is an analytic homeomorphism.

Some Preliminaries

In giving the proofs of the theorems which follow, we make use of concepts and results developed by the algebraic geometers. For convenience and clarity we have collected the information needed into six "facts" which are numbered and listed below. References [3] and [7] are good sources for the reader who wishes verification. We state these facts to apply to a real or complex variety V.

F1. The variety V has a minimal *field of definition, F,* which is an extension of the rationals such that V may be defined by polynomial equations with coefficients in F. There is no proper sub-field of F such that V may be defined by polynomials whose coefficients are in the sub-field. It is a theorem that F is unique for any given V.

F2. The *algebraic dimension,* dim(x), of a point $x = (x_1, \cdots, x_n)$ of V is the number of coordinates of x which one can select algebraically independent over F.

F3. A point x of V which has maximal dimension is called a *general point*; dim(V) is the dimension of its general points.

F4. Let x be a general point of algebraic dimension r. Then any set of r coordinates which takes on values at x that are algebraically independent over F serve to parametrize non-singularly a neighborhood of x in V. If V is a complex variety then this will be a complex parametrization.

F5. If x is a point of V and $\dim(x) < \dim(V)$, then x is a general point of a subvariety W, where $\dim(W) = \dim(x)$. W's field of definition is contained in V's.

F6. If a variety in E^n has algebraic dimension r then each set of $r + 1$ coordinates of E^n will satisfy a non-trivial polynomial on V.

THEOREM 5. *A representation $\mathfrak{B}$ of an algebraic manifold ($\mathfrak{M}$, $\mathfrak{R}$) in E^n is a sheet of an irreducible variety V whose dimension is the same as that of $\mathfrak{M}$.*

Proof. We first show that $\mathfrak{B}$ is contained in an irreducible variety, V, whose dimension does not exceed r, where r is the dimension of $\mathfrak{M}$.

From the definition of a representation, the coordinates $x_1, x_2,$ $\cdots x_n$, thought of as functions on $\mathfrak{B}$ (which we are here identifying with $\mathfrak{M}$) form a set of functions of $\mathfrak{R}$. Hence any $r + 1$ of them satisfy a non-trivial polynomial relation on $\mathfrak{B}$. Select one such relation for each $(r + 1)$-tuple of coordinates. And let V_0 be the variety which this set of equations defines.

V_0 may be reducible, in which case it is the union of two smaller varieties, say V_1 and V_2. The set $V_1 \cap \mathfrak{B}$ will be a closed set, so the subset $\mathfrak{S}$ of $\mathfrak{B}$ not in V_1 is an open subset of the manifold $\mathfrak{B}$. If $\mathfrak{S}$ is non-void, it is in V_1 so that every polynomial that vanishes on V_2 vanishes on $\mathfrak{S}$, and by analytic continuation, on all $\mathfrak{B}$. Thus if $\mathfrak{B}$ is not contained in V_1 it is contained in V_2.

Continuing this decomposition process, we must ultimately obtain an irreducible subvariety V of V_0 which contains $\mathfrak{B}$, since it is a theorem in algebraic geometry that there is no infinite sequence of varieties in which each variety is a proper subvariety of its predecessor.

Suppose $\dim(V) > r$; then there must be a neighborhood on V parametrized non-singularly by $r + k$ coordinates E^n (see **F4**). But this is inconsistent with the fact that any $r + 1$ of these coordinates must satisfy a polynomial relation on V_0, and hence also on the subvariety V. Therefore, $\dim(V) \leqq r$.

Now consider a point on $\mathfrak{B}$. Some r-tuple of the coordinates $x_1,$ $\cdots, x_n$ must suffice to parametrize without singularities a neighborhood

N_1 in $\mathfrak{B}$ of this point. For the r parametrizing coordinates we can find r numerical values which correspond to a point in N_1 and which are algebraically independent over V's field of definition. We can do this because a variety's field of definition must be enumerable and likewise the algebraic closure of any finite extension of it, while the field of real numbers has a higher cardinality. Thus we can obtain a point x in N_1, and therefore also on V, where $\dim(x) \geqq r$. Consequently $\dim(V) = \dim(x) = r$.

This general point x may be used to see that $\mathfrak{B}$ must be a sheet of V. From **F4**, a neighborhood N_2 of x in V is non-singularly parametrizable by the same r coordinates that parametrize N_1. Since x is in N_1, there must be a smaller neighborhood, N_3 of x that is contained both in N_1 and in N_2. Since N_3 is a neighborhood in V of x and contains no points of V that are not in $\mathfrak{B}$ we have verified condition (c) for a sheet.

Now suppose an analytic path joins x to another point of V and lies always in V. Near x it must lie in N_3 and hence must lie in $\mathfrak{B}$ for a finite interval of its length. But since $\mathfrak{B}$ is analytic this means the path can never leave $\mathfrak{B}$. So $\mathfrak{B}$ satisfies condition (b) for a sheet. And since any two points of an analytic manifold may be connected by an analytic arc (for example, any geodesic joining them) we have condition (a) and have shown that $\mathfrak{B}$ is a sheet of V.

THEOREM 6. *If $\mathfrak{B}$ is an analytic imbedding of a closed analytic manifold in E^n and is also a sheet of variety V, then $\mathfrak{B}$ is an algebraic representation of the manifold.*

Proof. Let r be the dimensionality of the manifold $\mathfrak{B}$. From the definition of a sheet, $\mathfrak{B}$ must have a point x which has a neighborhood that contains no points of V that are not also points of $\mathfrak{B}$. We also demand that $\dim(x)$ be at least r. The justifying argument is the same one we used in a similar situation while proving Theorem 5.

Now from **F5**, x is a general point of a subvariety W of V. As $\dim(x) \geqq r$, $\dim(W) \geqq r$. $\dim W$ cannot exceed r because if it did, W would have too high a dimensionality at its general point x [see **F4**] to fit into the above described neighborhood of x in V. Since W is non-singular and r-dimensional at x, it must coincide with $\mathfrak{B}$ near x. And since $\mathfrak{B}$ is an analytic manifold, it must be entirely contained in W.

Consider the ring $\mathfrak{R}$ of all single valued real analytic functions on $\mathfrak{B}$ which are algebraic functions of the coordinates of E^n. From **F6**, any $r + 1$ of the coordinate functions must satisfy a non-trivial polynomial relation on W, and hence on $\mathfrak{B}$. Since the other functions in $\mathfrak{R}$ depend algebraically on the coordinate functions any $r + 1$ of them will also satisfy a non-trivial polynomial on $\mathfrak{B}$.

It is now easy to see that the ring $\mathfrak{R}$ and the abstract analytic manifold of which $\mathfrak{B}$ is an imbedding will form an algebraic manifold $(\mathfrak{M}, \mathfrak{R})$ of which $\mathfrak{B}$ will be a representation.

THEOREM 7. *If two algebraic manifolds are equivalent as differentiable manifolds then they are also equivalent as algebraic manifolds.*

We must mention a few relevant facts and establish a minor lemma before giving the actual proof of Theorem 7. The facts to be mentioned are merely a few elementary properties of algebraic functions. We shall simply assert them; they can be verified by reference to texts on algebraic functions or elimination theory.

If a dependent variable is an algebraic function of n independent variables then the $n + 1$ variables (the independent variables and the dependent variable) will satisfy some non-trivial polynomial relation. The sum, difference, product, or quotient of algebraic functions is an algebraic function. An algebraic function of a set of algebraic functions of a set of independent variables is itself an algebraic function of the original independent variables. A derivative or partial derivative of an algebraic function is an algebraic function. Finally, if n variables $z_1, z_2, \cdots z_n$ depend algebraically on n other variables $x_1, x_2, \cdots x_n$ and there is an inverse functional relationship then the inverse functions $x_i(z_1, z_2, \cdots z_n)$ are algebraic functions. These are the facts about algebraic functions that we shall need in proving the theorem.

The minor lemma we spoke of above is just a specialization of Lemma 1. Return to the situation involved in Lemma 1 and reconsider the proof given, but now with the additional assumption that the imbedding functions $y_i(\lambda)$ can be chosen so as to be algebraic functions. The transformation $T : (z, \lambda) \rightarrow (x, \mu)$ will be algebraic since it was defined by operations which preserve the algebraic character of the original imbedding functions. Hence the inverse function $\lambda(x, \mu)$ will be algebraic and $y(x) = y(\lambda(x, 0))$ will also be algebraic.

This proves the result we need. Stated in words, it says that if a manifold is algebraically imbedded then the nearest-point function $y(x)$ will be algebraic. That is, each component $y_i(x_1, x_2, \cdots x_n)$ of the vector function $y(x)$ will be an algebraic function of the components $x_1, x_2, \cdots x_n$ of the vector x.

Proof of Theorem 7. Let $(\mathfrak{M}_1, \mathfrak{R}_1)$ and $(\mathfrak{M}_2, \mathfrak{R}_2)$ be the two algebraic manifolds and let $\mathfrak{B}_1$ and $\mathfrak{B}_2$ be representations of them in E^{n_1}, and E^{n_2}. By hypothesis $\mathfrak{B}_1$ and $\mathfrak{B}_2$ are differentiably homeomorphic. Let the function $\alpha(z)$ map $\mathfrak{B}_2$ into $\mathfrak{B}_1$ and let α be such that it sets up a both ways differentiable homeomorphism between $\mathfrak{B}_1$ and $\mathfrak{B}_2$. Here z stands for a point in E^{n_2} on $\mathfrak{B}_2$. Think of $\alpha(z)$ as a vector function taking values in E^{n_1}.

We begin by approximating the function $\alpha(z)$ by a function $\beta(z)$ whose components are polynomials in the coordinates of E^{n_2}. In addition, the first derivatives of β are to approximate the corresponding first derivatives of the C^1 vector function $\alpha(z)$.

β may be obtained by first approximating α by an analytic vector function constructed by any of several well known methods (e.g., smoothing by integration with a Gaussian kernel). The second step, approximating this analytic approximation by a vector function with polynomial coefficients, is covered by our Lemma 2.

Let x stand for a point in E^{n_1}. If x is near enough to $\mathfrak{B}_1$, the point $y(x)$ on $\mathfrak{B}_1$ which is nearest to x will be an analytic function of x. Moreover, $y(x)$ will be an algebraic function of x.

If the function β is a sufficiently good approximation to α, then for a point z on $\mathfrak{B}_2$ its image $\beta(z)$ will always be quite near $\mathfrak{B}_1$. Thus we may define a new function γ by setting $\gamma(z) = y(\beta(z))$; the function $\gamma(z)$ is defined for points on $\mathfrak{B}_2$. Now observe that applying the same operation to the original function α does not change it, i.e., $y(\alpha(z)) = \alpha(z)$, because α is defined only for points on $\mathfrak{B}_2$ and maps them into points on $\mathfrak{B}_1$. Therefore, since β approximates α, the new function γ also approximates α, for this is the same thing as saying that $y(\beta)$ approximates $y(\alpha)$.

If we choose local coordinate systems in $\mathfrak{B}_1$ and $\mathfrak{B}_2$, we may define locally the Jacobian of a C^1 mapping of $\mathfrak{B}_2$ into $\mathfrak{B}_1$. The Jacobian of the mapping defined by the function α must always be non-zero. Hence, if γ approximates α well enough, the Jacobian of the mapping that γ defines will also be nowhere vanishing. This must be so if β approximates α closely enough. Our conclusion here is that if β is a sufficiently good approximation to α then γ will define a both ways differentiable homeomorphism between $\mathfrak{B}_1$ and $\mathfrak{B}_2$.

Actually, γ defines an analytic homeomorphism, since γ was obtained as an analytic vector function of an analytic vector function. And for similar reasons, γ is algebraic, that is, the components of the vector function $\gamma(z)$ are algebraic functions of the components of z. Therefore the homeomorphism that γ defines between $\mathfrak{B}_1$ and $\mathfrak{B}_2$ must carry algebraic functions on either imbedded manifold into algebraic functions on the other.

From these observations on γ it is easy to see that the induced correspondence between algebraic functions on $\mathfrak{B}_1$ and algebraic functions on $\mathfrak{B}_2$ defines an isomorphism of the rings $\mathfrak{R}_1$ and $\mathfrak{R}_2$ (which may be thought of as rings of functions on $\mathfrak{B}_1$ and $\mathfrak{B}_2$). Thus $(\mathfrak{M}_1, \mathfrak{R}_1)$ and $(\mathfrak{M}_2, \mathfrak{R}_2)$ are equivalent.

Some Remarks

The concepts and results that appear in this paper suggest certain questions and conjectures which are perhaps sufficiently interesting to be mentioned.

Our results on the representation of manifolds by portions of algebraic varieties can probably be improved. Suppose a real variety happens to take the form of a compact analytic manifold, free of singularities. Then it would be what we call a *pure representation* of that manifold. It is probably true that given any closed differentiable manifold there is a variety forming a pure representation of the manifold. And it is plausible that our approximation theorem (Theorem 1) should also be true in a stronger form referring to approximation by pure representations.

The notion of a sheet of a variety also suggests a few questions. Does every variety have but a finite number of sheets? Are sheets always closed sets? Is each point of a variety in at least one of its sheets? Is condition (c) in the definition of a sheet superfluous?

While attempting to improve our results on algebraic representations of manifolds we were led to consider what we call rational manifolds. This concept differs from our concept of an algebraic manifold in that a distinguished class of functions is regarded as the class of rational (rather than algebraic functions).

We define a *real rational manifold* as a combination $(\mathfrak{M}, \mathfrak{F})$ of a closed analytic manifold $\mathfrak{M}$ and a field $\mathfrak{F}$ of functions on $\mathfrak{M}$ such that:

(1) Each function in $\mathfrak{F}$ is a real valued meromorphic function on $\mathfrak{M}$. By "meromorphic" we mean that such a function can be represented in the vicinity of any point of $\mathfrak{M}$ as the quotient of two single valued real analytic functions.
(2) There is a basic set on $\mathfrak{M}$ composed of functions in $\mathfrak{F}$.
(3) If $\mathfrak{M}$ is n-dimensional then there is a set S of n functions from $\mathfrak{F}$ such that every function in $\mathfrak{F}$ can be expressed as a rational function of the functions in S.

The conjecture is that any closed differentiable manifold can be made into a rational manifold. This would be a very powerful theorem if it could be proved. For example, it could be used to establish the conjecture we mentioned earlier about approximation by pure representations.

If a manifold $\mathfrak{M}$ were made into a rational manifold $(\mathfrak{M}, \mathfrak{F})$ and if an imbedding $\mathfrak{B}$ of $\mathfrak{M}$ in a Euclidean space were defined by a basic set of functions taken from $\mathfrak{F}$ then $\mathfrak{B}$ would be the real part of a rational variety, that is, a variety bi-rationally equivalent to a projective space.

Unfortunately, this conjecture on rational manifolds would probably not be easily established, if true, while a simple counterexample could demolish it. And there is not much evidence to support it.

We wish to express our gratitude to N. E. Steenrod, whose counsel was a great aid in the research leading to this paper. And we are indebted to the helpful advice and criticism given by D. C. Spencer and O. Zariski and the information gathered from conversations with several other individuals.

References

1. Hassler Whitney, *Differentiable Manifolds,* Ann. of Math., Vol. 37 (1936), pp. 645–80.
2. H. Seifert, *Algebraische Approximation von Mannigfaltigkeiten,* Math. Zeit., 41 (1936), pp. 1–17.
3. Oscar Zariski, *The Concept of a Simple Point of an Abstract Algebraic Variety,* Trans. Amer. Math. Soc., Vol. 62 (1947), pp. 1–52.
4. S. S. Cairns, *Normal Coordinates for Extremals Transversal to a Manifold,* Amer. J. of Math., Vol. 60 (1938), pp. 423–35.
5. H. Poincaré, *Analysis Situs,* Journal d l'Ecole Polytechnique, II Serie I, 1895.
6. Norman E. Steenrod, The Topology of Fibre Bundles, Princeton Mathematical Series, Vol. 14, Princeton University Press, Princeton, N.J., 1951.
7. André Weil, Foundations of Algebraic Geometry, Amer. Math. Soc. Colloquium Publications, Vol. 29, New York, N.Y., 1946.

CHAPTER 11

JOHN F. NASH, JR.

The Imbedding Problem
for Riemannian Manifolds

(Received October 29, 1954) (Revised August 20, 1955)

Introduction and Remarks

History. The abstract concept of a Riemannian manifold is the result of
an evolution in mathematical attitudes [1, 2]. In an earlier period mathematicians thought more concretely of surfaces in 3-space, of algebraic
varieties, and of the Lobatchevsky manifolds. As the more abstract view
of manifolds came into favor a question naturally arose: To what extent
are the abstract Riemannian manifolds a more general family than the
sub-manifolds of euclidean spaces?

This question has been considered in various specializations and
with assorted side conditions. In 1873 Schlaefli [3] discussed the local form of this imbedding problem. He conjectured that a neighborhood in an n-manifold would generally require an imbedding space of
$(n/2)(n + 1)$ dimensions. In 1901 Hilbert [4] obtained a negative result, showing that the Lobatshevsky plane is not realizable as a smooth
surface in E^3. Some contemporary negative theorems are due to Tompkins [5] and to Chern and Kuiper [6]. For example, a flat n-torus is not
realizable in less than $2n$ dimensions.

Janet [7] solved the local problem for two-manifolds with analytic metric in 1926, and Cartan [8] immediately extended the result to n-manifolds, treating it as an application of his theory of Pfaffian forms. The dimensionality requirement was $(n/2)(n+1)$, as conjectured by Schlaefli. This number is a plausible one, being the number of components of the metric tensor. The proof depended on power series development, so it was limited to local results and it required that the metric be analytic.

There are some theorems on the existence of isometric imbeddings in infinite dimensional spaces. This is a much simpler problem.

A recent discovery [9, 10] is that C^1 isometric imbeddings of Riemannian manifolds can be obtained in rather low dimensional spaces. At first glance some of these C^1 results seem inconsistent with the negative theorems, such as Hilbert's. Apparently C^1 imbeddings are very different from the smoother ones.

Until recently the only general results on imbeddings in the large were proved for the problem of Weyl. This problem is to realize in E^3 all two-manifolds with everywhere positive Gaussian curvature. Alexandrov [13] and Pogorelov [14] have been successful with a geometrical approach based on polyhedral approximations. H. Lewy [12] and L. Nirenberg [15] have treated the problem from the viewpoint of partial differential equations. These results can probably be sharpened with respect to differentiability, but dimension-wise they are clearly optimal.

Rigidity theory concerns the metric-preserving perturbations of an imbedding. A closed convex surface in E^3 is rigid, because it admits only trivial perturbations. But it becomes flexible if there is a hole in it. Apparently rigidity disappears completely when the imbedding space has enough dimensions.

Arrangement of this paper. There are four main divisions, called Parts A, B, C, and D. At the end of Part C the treatment of compact manifolds is complete and we state Theorem 2, which is essentially this: *Every compact Riemannian n-manifold is realizable as a sub-manifold of Euclidean $(n/2)(3n+11)$-space.* In Part D this is made to apply to non-compact manifolds by means of a device which reduces the non-compact problem to the compact case. The device is extravagant with dimensions. Theorem 3 realizes non-compact n-manifolds in $(n/2)(n+1)(3n+11)$ dimensions.

The core of this paper is in Part B. There a perturbation process is developed and applied to construct a small finite perturbation of an imbedding such that the perturbed imbedding induces a metric that differs by a specified small amount from the metric induced by the original imbedding. This work is summarized at the end of Part B in Theorem 1. The interesting thing about the perturbation process is that it does not seem special to this imbedding problem. It may be an illustration of a general method applicable to a variety of problems involving partial differential equations.

Part A is devoted to the fairly straightforward construction of a smoothing operator of the type required by the method of Part B. The operator's main properties are stated in equations (A15, 16, 17). In general, the four Parts are relatively independent in notation. Each depends only on the main results of the preceding part, not on the details.

Remarks. Some respects in which the results here should be improvable are these: The dimension bounds for the imbedding space should be lowered; the C^2 case should be included; and it should be proved that the process gives an analytic imbedding when the metric is analytic. The treatment of the C^2 and analytic cases would require new sets of estimates. A more unified approach to the problem which would not require the use of two separate sets of imbedding functions might reduce the dimension requirements substantially.

The methods used here may prove more fruitful than the results. Time will tell how much can be done with smoothing and "feed-back" methods like those applied in Part B. The device of Part D suggests an alternative way to imbed general two-manifolds by exploiting the results on Weyl's problem.

Acknowledgment. I am profoundly indebted to H. Federer, to whom may be traced most of the improvement over the first chaotic formulation of this work. N. Levinson also gave very helpful advice and information; and through constructive criticism, several others at M.I.T. helped me to improve the paper. This paper was supported in part by the Office of Naval Research.

JOHN F. NASH, JR.

The Imbedding Problem for Riemannian Manifolds

Part A: A General Smoothing Operator

This part develops an analytical tool, the smoothing operator, which is essential to the perturbation process developed in Part B. A smoothing operator is first constructed to act on real functions of n-variables, that is functions on E^n. Then we define a smoothing operator for a manifold $\mathfrak{M}$ by imbedding $\mathfrak{M}$ in E^n and extending scalar functions on $\mathfrak{M}$ to functions on E^n, which can be smoothed by the E^n operator. Finally, we devise a canonical representation for tensors on $\mathfrak{M}$ in terms of sets of scalars and use this to smooth tensors. We also obtain three important general inequalities that describe the action of smoothing on functions.

Smoothing Functions on E^n

Real functions of n-variables are smoothed by convolution with a certain kernel, K_θ, that we define below. Here θ is a parameter controlling the degree of smoothing. K_θ is defined by defining its Fourier transform $\bar{K}_\theta$.

Let $\psi(u)$ be a C^∞ function such that

$$\text{for } u \leqq 1: \qquad \psi(u) = 1,$$
$$\text{for } 1 \leqq u \leqq 2: \qquad \psi(u) \text{ is monotone decreasing,}$$
$$\text{for } u \geqq 2: \qquad \psi(u) = 0.$$

To illustrate, we could take $\psi = e^{e^{(1/1-u)}/u-2}$ in the range $1 < u < 2$.

Suppose $x_1, x_2, \cdots x_n$ are the coordinates of E^n and $\xi_1, \xi_2, \cdots, \xi_n$ are the corresponding coordinates of the Fourier transform space. We define the transform $\bar{K}_\theta$ of the kernel as

(A1)
$$\bar{K}_\theta = \psi(\xi/\theta), \text{ where}$$
$$\xi = \left(\xi_1^2 + \xi_2^2 + \cdots \xi_n^2\right)^{\frac{1}{2}}.$$

Thus $\bar{K}_\theta$ is a spherically symmetric non-negative C^∞ function which is 1 inside the sphere $\xi = \theta$, zero outside the sphere $\xi = 2\theta$, and smoothly decreasing with ξ in the annular region between the two spheres.

K_θ is the transform of $\bar{K}_\theta$; so K_θ is spherically symmetric; it is real because $\bar{K}_\theta$ is even; it is analytic because $\bar{K}_\theta$ vanishes, except for $\xi < 2\theta$; and $|K_\theta|$ decreases as rapidly as any negative power of the distance because all derivatives of $\bar{K}_\theta$ are continuous.

As θ varies K_θ will be more or less concentrated at the origin. But the integral over all of E^n of K_θ will always be the same. Since variation of θ changes $\tilde{K}_\theta$ only in a way corresponding to a change of scale in the transform space, K_θ will change in a similar way, with normalization being preserved. Specifically, we can relate K_θ to K_1 (where $\theta = 1$) by the equation

(A2) $\qquad K_\theta(x_1, x_2, \cdots x_n) = \theta^n K_1(\theta x_1, \theta x_2, \cdots \theta x_n).$

Effect of Convolution on Derivatives

Convolution and differentiation commute under favorable conditions:

(A3) $\qquad \dfrac{\partial}{\partial x_i}(K_\theta * f) = \left(\dfrac{\partial}{\partial x_i} K_\theta \right) * f.$

In our applications of this, f will vanish outside a compact region. Since K_θ (and also its derivatives, as we see below) decreases rapidly away from the origin, the favorable conditions will be more than met.

To consider a derivative of K_θ, observe that $\partial K_\theta / \partial x_i$, which we abbreviate to $K_{\theta,i}$ satisfies

$$\overline{K_{\theta,i}} = \xi_i(-1)^{\frac{1}{2}} \tilde{K}_\theta.$$

Thus the transform of $K_{\theta,i}$ is a C^∞ function and vanishes outside the sphere $\xi = 2\theta$. So $|K_{\theta,i}|$ decreases rapidly in the same manner as $|K_\theta|$, and the same is true for the higher derivatives of K_θ.

From (A2) we see that

$$K_{\theta,i}(x_1, \cdots, x_n) = \theta^{n+1} K_{1,i}(\theta x_1, \cdots, \theta x_n),$$

so (A3) can be written as

$$
\begin{aligned}
(K_\theta * f)_{,i} &= \theta^{n+1} K_{1,i}(\theta x_1, \cdots, \theta x_n) * f \\
&= \theta^{n+1} \int \cdots \int K_{1,i}(\theta y_1, \cdots, \theta y_n) \\
&\qquad\qquad\qquad f(x_1 - y_1, \cdots, x_n - y_n)\, dy_1 \cdots dy_n \\
&= \theta \int \cdots \int K_{1,i}(z_1, \cdots, z_n) \\
&\qquad\qquad f((x_1 - z_1/\theta), \cdots, (x_n - z_n/\theta))\, dz_1 \cdots dz_n.
\end{aligned}
$$

Here we considered the integral formula for convolution, letting the kernel carry the dummy variables $y_1, y_2, \cdots, y_n$. Then we made a change of variables: $\theta y_i = z_i$. Now using the last equation we can say

$$\left| (K_\theta * f)_{,i} \right| \leqq \theta \left(\max |f| \right) \int \cdots \int \left| K_{1,i} \right| \, dz_1 \cdots dz_n$$
$$\leqq C \, \theta \max |f|,$$

where C is the integral of the absolute value of $K_{1,i}$. Similarly we could obtain a bound for the size of a higher derivative of $K_\theta * f$. Each such bound would involve a constant, analogous to C, and θ to the power equal to the order of the derivative. A higher order derivative can be considered a derivative of a lower order derivative and we can bound it in terms of the maximum size of that derivative, before convolution with K_θ. To illustrate,

$$(K_\theta * f)_{,ji} = (K_\theta * f_{,j})_{,i} = (K_{\theta,i} * f_{,j}),$$
$$\therefore \quad \left| (K_\theta * f)_{,ji} \right| \leqq C \, \theta \max \left| f_{,j} \right|.$$

If we are not concerned with the precise sizes of the constants, such as C, which appear in these estimates we can put them all in one comprehensive statement. Let $\max^{(s)} f$ stand for the maximum of the values attained by the absolute values of all derivatives of f of order s at all points of E^n.

Then we can say generally:

(A4) $\max^{(r)} (K_\theta * f) \leqq C_{rs} \theta^{r-s} \max^{(s)} f,$ when $r \geqq s$.

Here C_{rs} is to be a constant, independent of f, and in fact depending only on $r - s$.

Effect of Varying θ

We need to know how rapidly $K_\theta * f$ and its derivatives change as θ varies. Of course we can say

$$\frac{\partial}{\partial \theta} (K_\theta * f) = \frac{\partial K_\theta}{\partial \theta} * f,$$

since we do not think of f as varying with θ. $\partial K_\theta / \partial \theta$ will also be a kernel with good properties, as we see below.

We can begin by considering

157

$$
\overline{\frac{\partial}{\partial \theta} K_\theta} = \frac{\partial}{\partial \theta} \bar{K}_\theta = \frac{\partial}{\partial \theta} \left[\psi(\xi/\theta) \right] = -\xi/\theta^2 \psi'(\xi/\theta)
$$

(A5)

$$
= \theta^{-1} \chi(\xi/\theta).
$$

Here we introduce a new function, χ, defined by $\chi(u) = -u\psi'(u)$. Observe that $\chi(u) = 0$ for $u \leq 1$ or $u \geq 2$, and that $\chi(u) > 0$ for $1 < u < 2$. Also χ is C^∞, so we see that $\partial K_\theta/\partial \theta$ has the same general properties of analyticity, smallness at infinity, etc., that K_θ has.

Let L stand for the value of $\partial K_\theta/\partial \theta$ at $\theta = 1$. We can express $\partial K_\theta/\partial \theta$ in terms of L in a manner analogous to the way we obtained in (A2) an expression for K_θ as $\theta^n K_1(\theta x_1, \cdots)$. The only difference in the situations is the appearance of θ^{-1} in (A5). So we obtain

(A6)
$$
\frac{\partial}{\partial \theta} K_\theta = \theta^{n-1} L(\theta x_1, \theta x_2, \cdots, \theta x_n).
$$

We want to express $L(x_1, x_2, \cdots, x_n)$ as the sum of n special functions L_i. We do this via the transform of L, which is $\chi(\xi)$. Thus we shall have

(A7)
$$
\bar{L} = \chi(\xi) = \sum_i \bar{L}_i.
$$

To define the $\bar{L}_i$ we construct a non-negative C^∞ function α_i for each transform variable ξ_i:

$$
\alpha_i = 0, \quad \text{for} \quad |\xi_i| \leq (2n)^{-\frac{1}{2}},
$$

and

$$
\alpha_i = e^{((2n)^{-\frac{1}{2}} - |\xi_i|)^{-1}}
$$

for $|\xi_i| > (2n)^{-\frac{1}{2}}$.

Observe that when $\xi > (2)^{-\frac{1}{2}}$ then $\xi_1^2 + \xi_2^2 + \cdots + \xi_n^2 > \frac{1}{2}$ and some $\xi_i > (2n)^{-\frac{1}{2}}$ so that some $\alpha_i > 0$. Therefore $\alpha_1 + \alpha_2 + \cdots + \alpha_n$ is positive and C^∞ in the region $1 \leq \xi \leq 2$ where $\chi(\xi)$ is non-vanishing. So we can define

$$
\bar{L}_i = \frac{\alpha_i}{\alpha_1 + \alpha_2 + \cdots + \alpha_n} \chi(\xi), \quad \text{for} \quad \xi \geq 1, \quad \text{and}
$$
$$
\bar{L}_i = 0, \quad \text{for} \quad \xi < 1.
$$

These functions will be C^∞ everywhere and will satisfy (A7). Each $\bar{L}_i$ has the important properties:

(a) $\bar{L}_i = 0$ for $\xi \geq 2$
(b) $\bar{L}_i = 0$ for $|\xi_i| \leq (2n)^{-\frac{1}{2}}$.

The corresponding kernels L_i will clearly have all the good behavior properties at infinity, etc., that K_θ has. And we have

$$\sum_i L_i = L = \left\{ \begin{array}{l} \text{the value at } \theta = 1 \\ \text{of } \dfrac{\partial}{\partial \theta} K_\theta \end{array} \right\}.$$

Purpose of the L_i Kernels

Each L_i is so constructed that when one forms the indefinite integral $\int_{-\infty}^{x_i} L_i \, dx_i$ the result is still a function that is small at infinity. We see this again via the transform $\bar{L}_i$. Let $\bar{H}_i^r$ stand for the rth member of a series of kernels developed from L_i and defined by

$$\bar{H}_i^r = \left(\xi_i (-1)^{\frac{1}{2}} \right)^{-r} \bar{L}_i.$$

The properties of $\bar{L}_i$ that were emphasized above insure that $\bar{H}_i^r$ will be a C^∞ function of $\xi_1, \xi_2, \cdots, \xi_n$ which vanishes when $|\xi_i| \leq (2n)^{-\frac{1}{2}}$ or $|\xi| \geq 2$. Therefore the kernels H_i^r are analytic functions which decrease rapidly away from the origin.

The important property of the H_i^r is that

$$\frac{\partial^r H_i^r}{\partial x_i^r} = L_i. \quad \text{Also} \quad H_i^{r+1} = \int_{-\infty}^{x_i} H_i^r \, dx_i,$$

because

$$\int_{-\infty}^{\infty} H_i^r \, dx_i = 0, \quad \text{since} \quad \bar{H}_i^r = 0 \quad \text{when} \quad \xi_i = 0.$$

The H_i^r help us to estimate the size of $\partial K_\theta / \partial \theta * f$ from data on the size of derivatives of f. By (A6) we have

$$\frac{\partial}{\partial \theta} K_\theta * f = \left\{ \theta^{n-1} \sum_i L_i(\theta x_1, \cdots \theta x_n) \right\} * f(x_1, \cdots, x_n).$$

Changing variables, $\theta x_i \to x_i$, as we did when estimating $(K_\theta * f)_{,i}$, we obtain

$$\partial K_\theta / \partial \theta * f = \theta^{-1} \left[\left\{ \sum_i L_i(x_1, \cdots, x_n) \right\} * f(x_1/\theta, \cdots, x_n/\theta) \right]$$

$$= \theta^{-1} \sum_i \left\{ L_i(x_1, \cdots, x_n) * f(x_1/\theta, \cdots, x_n/\theta) \right\}$$

$$= \theta^{-1} \sum_i \left\{ H_i^r(x_1, \cdots, x_n) * \frac{\partial^r}{\partial x_i^r} [f(x_1/\theta, \cdots, x_n/\theta)] \right\}$$

$$= \theta^{-1} \sum_i \left\{ H_i^r(x_1, \cdots, x_n) * \theta^{-r} \frac{\partial^r}{\partial (x_i/\theta)^r} f(x_1/\theta, \cdots x_n/\theta) \right\}$$

$$= \theta^{-r-1} \sum_i \left\{ H_i^r * \frac{\partial^r}{\partial (x_i/\theta)^r} f(x_1/\theta, \cdots, x_n/\theta) \right\} . \qquad \textbf{159}$$

By the series of equations above we have expressed $(\partial K_\theta / \partial \theta) * f$ entirely in terms of rth order derivatives of f. These equations are, of course, invalid unless the rth order derivatives of f are continuous, which we assume. They lead to an inequality:

$$\left| \frac{\partial}{\partial \theta} K_\theta * f \right| \le \theta^{-r-1} \max \left\{ \sum_i \left| \frac{\partial^r f}{\partial x_i^r} \right| \right\} \max_i \int \cdots \int |H_i^r| \, dx_1 \cdots dx_n$$

$$\le C_r \theta^{-r-1} \max^{(r)} f .$$

Since a derivative of $(\partial K_\theta / \partial \theta) * f$ is the same as $(\partial K_\theta / \partial \theta)$ convoluted with the corresponding derivative of f, we can also say

$$\left| \left(\frac{\partial}{\partial \theta} K_\theta * f \right)_{,i} \right| \le C_r \theta^{-r-1} \max^{(r)}(f, i)$$

$$\le C_r \theta^{-r-1} \max^{(r+1)} f .$$

Extending this principle, we obtain

$$\max^{(s)} \left[\frac{\partial}{\partial \theta} K_\theta * f \right] \le C_r \theta^{-r-1} \max^{(r+s)} f ,$$

(A8)   or

$$\le C_{t-s} \theta^{s-t-1} \max^{(t)} f .$$

This is valid only for $t \ge s$, so far as we have shown.

Actually (A8) is also valid for $t < s$, if appropriate constants C_r are used for the negative r values. We don't need the L_i to see this. Beginning with (A6), we obtain

$$\left(\frac{\partial}{\partial\theta}K_\theta\right)*f = \theta^{n-1}L(\theta x_1,\cdots,\theta x_n)*f(x_1,\cdots,x_n),$$

$$\therefore\quad \frac{\partial^r}{\partial x_i^r}\left[\frac{\partial}{\partial\theta}K_\theta*f\right] = \theta^{r+n-1}\frac{\partial^r}{\partial(\theta x_i)^r}L(\theta x_1,\cdots,\theta x_n)*f(x_1,\cdots,x_n).$$

Making the change of variables $\theta x_i \to x_i$, the right hand term is

$$\theta^{r-1}\frac{\partial^r}{\partial x_i^r}L(x_1,\cdots,x_n)*f(x_1/\theta,\cdots,x_n/\theta), \leq C_{-r}\theta^{r-1}\max|f|.$$

Again we can generalize by replacing f by $(\partial^s/\partial x_i^s)f$. And we could deal with mixed partial derivatives. The general result would be

$$\max{}^{(s)}\left[\frac{\partial}{\partial\theta}K_\theta*f\right] \leq C_{-r}\theta^{r-1}\max{}^{s-r}f.$$

This corresponds to (A8) with $-r$ in place of r. So we see that (A8) holds with positive or negative r or with $s \leq t$ or $s > t$.

Smoothing on a Manifold

Let the manifold $\mathfrak{M}$ be compact and analytic in the strong sense, so that it has an analytic imbedding $\mathfrak{R}$ in a euclidean space E^n. We can find a surrounding neighborhood, $\mathfrak{R}$, in E^n of $\mathfrak{R}$; and $\mathfrak{R}$ can be such that for any point x in $\mathfrak{R}$ there is a unique point $y(x)$ on $\mathfrak{R}$ which is the point of $\mathfrak{R}$ that is nearest to x. Also $y(x)$ can be an analytic function throughout $\mathfrak{R}$.[1]

Now let

$$\varphi(x) = \psi\left(\frac{\text{distance from } x \text{ to } y(x)}{\varepsilon}\right)$$

where ψ is the C^∞ function defined before above (A1) and ε is a small constant. If we assume ε is sufficiently small, $\varphi(x)$ will be a C^∞ function throughout $\mathfrak{R}$ and will vanish at all points near the boundary of $\mathfrak{R}$. Assume also that the definition of $\varphi(x)$ is extended by making it vanish identically outside $\mathfrak{R}$. Then $\varphi(x)$ is C^∞ everywhere.

---

1. The existence of such a neighborhood $\mathfrak{R}$ is shown in Lemma 1 of [16].

Now if $f(y)$ is a function defined on $\Re$ we define an extension $f(x)$ to E^n by putting

$$f(x) = \varphi(x) f(y(x)), \quad \text{for} \quad x \in \Re,$$

and

$$f(x) = 0, \quad \text{for} \quad x \notin \Re.$$

This extends $f(y)$ to a function $f(x)$ which agrees with the original function on $\Re$ and has the same degree of differentiability.

The method of smoothing is simple. Beginning with $f(y)$ on $\Re$ one extends to $f(x)$ in E^n. Then $K_\theta * f(x)$ is the smoothed function. The final (a logical formality) step is to restrict the definition of $K_\theta * f(x)$ to $\Re$ and again have a function defined only on $\Re$. But we must do more than simply present this definition of smoothing on $\Re$; we need to know how it affects derivatives of the function smoothed, etc. To do this we need the following concept.

A Standard Size Concept for Derivatives

$\Re$ has an analytic metric induced by the imbedding in E^n. So at each point p of $\Re$ we can set up an internal system of geodesic normal coordinates. This system is not unique, but is unique up to orthogonal transformations. At p we can measure the size of the derivatives of order r of a function by considering all the systems of geodesic normal coordinates at p. We define $\text{size}_p^{(r)} f$ as the maximum over all these systems of the maximum of the absolute values of the various rth order derivatives of f with respect to the coordinates of that system. Then we call $\text{size}^{(r)} f$ the maximum of $\text{size}_p^{(r)} f$ over all points p of $\Re$.

We need to know how the measure $\text{size}^{(r)} f(y)$ of the sizes of the derivatives of f as a function on $\Re$ is related to the measure $\max^{(r)} f(x)$ of the sizes of the derivatives of the function extended to E^n. And when $f(y)$ is obtained by restricting the range of definition of a function defined throughout E^n we need to know how the sizes of the internal derivatives will be related to the sizes of the derivatives with respect to coordinates of E^n. In the first case it is fairly easy to see that there will be general inequalities of the form

(A9) $$\max{}^{(r)}f(x) \leqq \sum_{k=0}^{r} B_k^r \, \text{size}^{(k)}f(y)$$

where the coefficients are constants determined by the imbedding of $\mathfrak{R}$ in E^n and the function φ which was used in the extension of the function $f(y)$ defined only on $\mathfrak{R}$ to the function $f(x)$ defined on E^n.

Similarly, when a function $g(x)$ defined on E^n is specialized to a function, say $g(y)$, defined only on $\mathfrak{R}$ there is a conversion from bounds on derivatives with respect to the E^n coordinates $(\max^{(r)})$ to the internal measure of the size of derivatives. This has the form

(A10) $$\text{size}^{\,(r)}g(y) \leqq \sum_{k=0}^{r} D_k^r \max{}^{(k)}g(x).$$

Actually, $D_0^r = 0$, except for the trivial case $r = 0$, where $D_0^0 = 1$. The constants D_s^r depend only on the imbedding $\mathfrak{R}$.

Effect of Smoothing on a Manifold on Derivatives

We are now ready to see how smoothing of a function on a manifold acts on derivatives, etc., and relate this action to the size of the original function and its derivatives. Suppose $f(y)$ was the original function. Then smoothing proceeds thus:

(a) $\quad f(y) \rightarrow f(x) \quad$ by extension to E^n,

(b) $\quad f(x) \rightarrow g(x) = K_\theta * f(x)$,

(c) $\quad g(x) \rightarrow g(y) \quad$ by restriction to $\mathfrak{M}$.

We call $g(y) = S_\theta f(y)$ so that here S_θ stands for the total operation of smoothing.

Corresponding to the two general inequalities, (A4) and (A8), for smoothing in E^n, we obtain two general inequalities for smoothing on $\mathfrak{R}$. For example, (A9) gives us bounds on $\max^{(r)} f(x)$ from $\text{size}^{(r)}$ data on $f(y)$. Then (A4) gives us $\max^{(r)}$ data on $g(x)$ from this data on $f(x)$. Finally (A10) gives us $\text{size}^{(r)}$ data on $g(y)$ from the $\max^{(r)} g(x)$ data. The outcome is a bound on $\text{size}^{(r)}g(y)$ in terms of $\text{size}^{(s)}f(y)$, $\text{size}^{(s-1)}f(y)$, $\cdots$, $\text{size}^{(0)}f(y)$. If we use $S_\theta f$ for $g(y)$ and f for $f(y)$, and if we weaken the form of the bound by using the maximum constant involved, we get a bound of the form

(A11) $\quad \text{size}^{(r)}\left[S_\theta f\right] \leqq H_{rs}\theta^{r-s}\sum_{t=0}^{s} \text{size}^{(t)}f, \quad \text{for } \theta \geqq 1, \, r \geqq s.$

Exactly analogously we obtain from (A8):

$$\text{(A12)} \qquad \text{size}^{(r)}\left[\frac{\partial}{\partial\theta}S_\theta f\right] \le J_{rs}\theta^{r-s-1}\sum_{t=0}^{s}\text{size}^{(t)}f, \quad \text{for } \theta \ge 1.$$

We use the restriction $\theta \ge 1$ so that we can majorize lower powers of θ by θ^{r-s} or θ^{r-s-1} in the two inequalities. The H_{rs} and J_{rs} coefficients depend only on the imbedding $\mathfrak{R}$ and φ.

Smoothing of Tensors

Here the first step is to express each tensor in a normalized (non-tensorial) form in terms of a set of scalar functions defined over all of $\mathfrak{R}$. This corresponds to using a specific redundant coordinate system on $\mathfrak{R}$ in order to have a coordinate system without singularities. Let $x^1, x^2, \cdots, x^n$ be the coordinates of E^n and let $u^1, u^2, \cdots, u^\nu$ be any local system of coordinates in $\mathfrak{R}$, such as one of the family of geodesic normal coordinate systems used in defining the $\text{size}^{(r)}$ concept.

The imbedding defines a transformation

$$\text{(A13)} \qquad \left(u^1, u^2, \cdots, u^\nu\right) \to \left(x^1, x^2, \cdots, x^n\right)$$

and the nearest point function $y(x)$ defines a transformation $x \to y(x)$ in the neighborhood $\mathfrak{R}$, which gives a transformation of coordinates

$$\text{(A14)} \qquad \left(x^1, x^2, \cdots, x^n\right) \to \left(u^1, u^2, \cdots, u^\nu\right).$$

Both transformations are analytic and their composition in either order is the identity on $\mathfrak{R}$.

Now suppose $T^{\alpha\beta\cdots}_{\gamma\delta\cdots}$ is a tensor on $\mathfrak{R}$ referred to the coordinates $u^1, u^2, \cdots, u^\nu$. Then we define

$$\mathfrak{T}^{ij\cdots}_{kl\cdots} = T^{\alpha\beta\cdots}_{\gamma\delta\cdots}\frac{\partial x^i}{\partial u^\alpha}\frac{\partial x^j}{\partial u^\beta}\cdots\frac{\partial u^\gamma}{\partial x^k}\frac{\partial u^\delta}{\partial x^l}\cdots,$$

with the summation convention operating, and with $\partial x^i/\partial u^\alpha$, etc., taken from (A14). This definition has the correct invariance properties, so $\mathfrak{T}^{ij\cdots}_{kl\cdots}$ is defined globally on $\mathfrak{R}$ and is completely independent of the coordinates $u^1, u^2, \cdots, u^\nu$ through which it is obtained.

Because the composition of (A13) and (A14) is the identity, it is easy to see that the reverse conversion

$$T^{\alpha\beta\cdots}_{\gamma\delta\cdots} = \mathfrak{T}^{ij\cdots}_{kl\cdots}\frac{\partial u^\alpha}{\partial x^i}\frac{\partial u^\beta}{\partial x^j}\cdots\frac{\partial x^k}{\partial u^\gamma}\frac{\partial x^l}{\partial u^\delta}\cdots$$

yields the original tensor again from the normalized form $\mathfrak{T}^{ij\cdots}_{kl\cdots}$.

The smoothing operation for a tensor consists of three steps: (a) conversion to normalized form, (b) smoothing of each component of the normalized form by S_θ, and (c) converting the result back to a tensor on $\mathfrak{R}$ via the reverse conversion.

Derivative Size Concept for Tensors

If T is a tensor we consider the standard local coordinate systems defined before and, at the center of each system, we consider all the derivatives of order r of each component of T. The maximum of the absolute values of these is then the local size$^{(r)}$ of T. Then the maximum of this over all the standard systems is size$^{(r)} T$.

Associated with the conversion of tensors to normalized form and with the reverse conversion there will be conversions from size$^{(r)} T$ measures to size$^{(r)}$ measures on the components of $\mathfrak{T}$. These will give us inequalities quite analogous to (A9) and (A10). By combining these inequalities (we shall not bother to write them out) with (A11) and (A12) to estimate the effect of S_θ on the components of $\mathfrak{T}$, we can obtain the analogous bounds for smoothing of tensors via the process $T \to \mathfrak{T} \to S_\theta \mathfrak{T} \to S_\theta T$ (definition of $S_\theta T$). These bounds are

(A15)
$$\text{size}^{(r)}(S_\theta T) \leq L_{rs}\theta^{r-s}\sum_{t=0}^{s} \text{size}^{(t)} T$$
$$\text{for } r \geq s, \theta \geq 1$$

and

(A16) $\quad \text{size}^{(r)}\left[\left(\dfrac{\partial}{\partial\theta} S_\theta\right) T\right] \leq M_{rs}\theta^{r-s-1}\sum_{t=0}^{s}\text{size}^{(t)} T \quad$ for $\theta \geq 1$.

Note that the process for smoothing tensors preserves certain simple properties that a tensor may have, such as symmetry or skew-symmetry, etc.

Concluding Remarks

The size$^{(r)}$ concept is suited to routine estimations concerning sums, differences, or products of functions or tensors. For example,

$$\text{size}^{(1)}(fg) \leq \left(\text{size}^{(0)}f\right)\left(\text{size}^{(1)}g\right) + \left(\text{size}^{(1)}f\right)\left(\text{size}^{(0)}g\right),$$
$$\text{size}^{(0)}\left(T_i^j S_j^k\right) \leq v\, \text{size}^{(0)}\left(T_i^j\right) \text{size}^{(0)}\left(S_j^k\right).$$

In the second estimate we invoke the summation convention, and v should be the dimensionality of $\mathfrak{R}$. These remarks are made to prepare for the frequent use of such elementary estimations in Part B, where we shall not specifically mention these properties of $\text{size}^{(r)}$ in the instances where they are used.

Another general bound referring to the action of S_θ is derivable from (A15) and (A16). This is

(A17) $\quad \text{size}^{(r)}(T - S_\theta T) \leqq N_{rs}\theta^{r-s}\displaystyle\sum_{t=0}^{s} \text{size}^{(t)}T \text{ for } s \geqq r, \theta \geqq 1.$

In the case $s = r$ this is derived trivially from (A15). When $s > r$ we use

$$T - S_{\theta_1} T = \int_{\theta_1}^{\infty} \left(\frac{\partial}{\partial\theta} S_\theta\right) T \, d\theta$$

and apply (A16).

Our more or less elaborate development of S_θ was undertaken to give us a smoothing operator for which (A15), (A16), and (A17) would hold. Without care in the definition one would obtain a weaker set of bounds. (A17) would probably have $\theta^{\max(r-s,-2)}$ instead of θ^{r-s}. (A15) would probably not be weakened.

There is a good heuristic interpretation for an operator such as S_θ. It is a low-pass filter which passes undiminished all frequencies below θ. Above 2θ it cuts off completely. Between θ and 2θ there is a variable attenuation, decreasing with increasing frequency so that the characteristic of the filter for all frequencies is a C^∞ function of frequency.

Part B: The Metric Perturbation Theorem

The Perturbation Device

The perturbation process developed in this part of the paper is based on a method for finding an infinitesimal change in the imbedding of a manifold that will effect a specified infinitesimal change in the metric induced by that imbedding. The smoothing operator S_θ of Part A is used in connection with this method, the "perturbation device."

Let $\mathfrak{M}$ be a compact n-manifold smoothly imbedded in E^m. Let the Cartesian coordinates of E^m be $z_1, z_2, \cdots, z_m$. Referred to a set

$x_1, x_2, \cdots, x_n$ of local coordinates in $\mathfrak{M}$, the metric tensor induced by the imbedding is

$$(B1) \qquad g_{ij} = \sum_\alpha \frac{\partial z_\alpha}{\partial x_i} \frac{\partial z_\alpha}{\partial x_j}.$$

We can consider perturbations of the imbedding as rates of change, measured with respect to the change of a parameter. The parameter is unspecified and we indicate the rate of change of any quantity by placing a dot over it. Thus we have

166

$$(B2) \qquad \dot{g}_{ij} = \sum_\alpha \frac{\partial z_\alpha}{\partial x_i} \frac{\partial \dot{z}_\alpha}{\partial x_j} + \sum_\alpha \frac{\partial \dot{z}_\alpha}{\partial x_i} \frac{\partial z_\alpha}{\partial x_j},$$

which follows from (B1).

We want a method by which we can determine $\{\dot{z}_\alpha\}$ satisfying (B2) when $\{\dot{g}_{ij}\}$ is specified. We can make this problem simpler to solve by adding another condition to be satisfied by the rate $\{\dot{z}_\alpha\}$ of change of the imbedding. This is

$$(B3) \qquad \sum_\alpha \frac{\partial z_\alpha}{\partial x_i} \dot{z}_\alpha = 0, \quad \text{for all } i,$$

which requires the perturbation $\{\dot{z}_\alpha\}$ to be normal to the imbedding.

Consider the result of differentiating (B3) with respect to x_j:

$$(B4) \qquad \sum_\alpha \frac{\partial z_\alpha}{\partial x_i} \frac{\partial \dot{z}_\alpha}{\partial x_j} = -\sum_\alpha \frac{\partial^2 z_\alpha}{\partial x_j \partial x_i} \dot{z}_\alpha.$$

The left member occurs in (B2) and the right member is symmetric in i and j (since the imbedding is to be reasonably smooth). Consequently we can use (B4) to modify (B2) and obtain

$$(B5) \qquad \dot{g}_{ij} = -2 \sum_\alpha \frac{\partial^2 z_\alpha}{\partial x_j \partial x_i} \dot{z}_\alpha.$$

This is the condition the perturbation $\{\dot{z}_\alpha\}$ should satisfy to effect the metric perturbation $\{\dot{g}_{ij}\}$ when (B3) holds.

Now we have a much simpler type of requirements relating $\{\dot{z}_\alpha\}$ to $\{\dot{g}_{ij}\}$. Together (B3) and (B5) form a system of linear equations to be satisfied by the $\dot{z}_\alpha$ whereas before we had partial differential equations in the $\dot{z}_\alpha$.

How and when can we solve (B3)–(B5) for $\{\dot{z}_\alpha\}$ after $\{\dot{g}_{ij}\}$ has been specified? The number of variables $\dot{z}_\alpha$, which is m, should be at least as large as the number of linear equations, which is $\frac{1}{2}n^2 + 1\frac{1}{2}n$, taking the i, j symmetry into account. Probably m will have to be larger than $\frac{1}{2}n^2 + 1\frac{1}{2}n$ to insure that the equations are not singular at some points of $\mathfrak{M}$. So the equations will probably be underdetermined.

In Part C we construct an imbedding of $\mathfrak{M}$ such that (B3)–(B5) is everywhere non-singular. Here in Part B we assume that the imbedding has this property and simply make our results conditional upon this.

We must find a way to select a particular solution $\{\dot{z}_\alpha\}$, and do this in a smooth way, when the equations are underdetermined. A very simple requirement,

$$(B6) \qquad \sum_\alpha (\dot{z}_\alpha)^2 = \text{minimum},$$

subject to satisfaction of (B3–B5),

selects a particular solution in a satisfactory manner. $\{\dot{z}_\alpha\}$ will have the same degree of differentiability as $\{\dot{g}_{ij}\}$.

The geometrical interpretation of the effect of (B6) is that it selects the nearest point to the origin in the plane of solutions $\{\dot{z}_\alpha\}$ of (B3)–(B5). We can also study the effect of (B6) from a more formal viewpoint. The system (B3)–(B5) is of the form:

(a) $$\sum_{\alpha=1}^m C_{\mu\alpha} \dot{z}_\alpha = \varphi_\mu.$$

If we assume a solution in the form

(b) $$\dot{z}_\alpha = \sum_{v=1}^{v=\frac{1}{2}n^2+1\frac{1}{2}n} C_{v\alpha} d_v,$$

then the d_v's must satisfy

(c) $$\sum_{\alpha, v} C_{\mu\alpha} C_{v\alpha} d_v = \varphi_\mu.$$

Let

(d) $$E_{\mu v} = \sum_\alpha C_{\mu\alpha} C_{v\alpha},$$

then (c) becomes

(e) $$\sum_v E_{\mu v} d_v = \varphi_\mu.$$

This last equation (e), will be non-singular if $\det \|E_{\mu v}\|$ is not zero. However, this is Gram's determinant for the matrix $\|C_{\mu\alpha}\|$ and it cannot vanish unless $\|C_{\mu\alpha}\|$ has less than maximal rank $(\frac{1}{2}n^2 + 1\frac{1}{2}n)$. Because we are assuming that (a), which is (B3)–(B5) in a condensed notation, is non-singular, we know that rank $\|C_{\mu\alpha}\| = \frac{1}{2}n^2 + 1\frac{1}{2}n$. So (e) is non-singular. Because (e) is not underdetermined it has a solution in the form

(f) $$d_\nu = \|E_{\mu\nu}\|^{-1} \cdot \{\varphi_\mu\}.$$

Now from (b) we can express a special solution $\{\dot{z}_\alpha^*\}$ of (a) (or of (B3)–(B5), which is the same thing) in the form

(g) $$\dot{z}_\alpha^* = \|C_{\nu\alpha}\| \cdot \|E_{\mu\nu}\|^{-1} \cdot \{\varphi_\mu\}.$$

This special solution of (a) happens to be the one for which $\sum_\alpha (\dot{z}_\alpha)^2$ is minimized. Suppose $\{\dot{z}_\alpha\}$ is any other solution of (a). Then

(h) $$\sum_\alpha C_{\mu\alpha}(\dot{z}_\alpha - \dot{z}_\alpha^*) = 0.$$

We can write

(i) $$\sum_\alpha (\dot{z}_\alpha)^2 - \sum_\alpha (\dot{z}_\alpha^*)^2 = \sum_\alpha (\dot{z}_\alpha - \dot{z}_\alpha^*)^2 + 2\sum_\alpha \dot{z}_\alpha^*(\dot{z}_\alpha - \dot{z}_\alpha^*),$$

and the last term vanishes because

(j) $$\begin{aligned}
2\sum_\alpha \dot{z}_\alpha^*(\dot{z}_\alpha - \dot{z}_\alpha^*) &= 2\sum_\alpha \left[\sum_\nu d_\nu C_{\nu\alpha}\right](\dot{z}_\alpha - \dot{z}_\alpha^*) \\
&= 2\sum_\nu d_\nu \sum_\alpha C_{\nu\alpha}(\dot{z}_\alpha - \dot{z}_\alpha^*) \\
&= 2\sum_\nu d_\nu \cdot 0 \\
&= 0.
\end{aligned}$$

Here we employed the expression (b) for $\dot{z}_\alpha^*$.

Now that we know the last term of (i) is zero it follows that the right hand side is positive and $\sum_\alpha (\dot{z}_\alpha)^2 > \sum_\alpha (\dot{z}_\alpha^*)$. Thus the special solution of (B3)–(B5) that is given by (g) is the same as the one selected by (B6). This shows that (B6) determines a solution which is a well behaved function of $\{\dot{g}_{ij}\}$ and the derivatives of the imbedding functions, so long as (B3)–(B5) remains a non-singular system.

The solution of (B3)–(B5) determined by (B6), or equivalently by (g), has the following form of linear dependence on $\{\dot{g}_{ij}\}$ and analytic dependence on the imbedding derivatives:

(B7) $$\dot{z}_\alpha = \sum_{i \le j} \dot{g}_{ij} F_{\alpha ij} \left(\left\{ \frac{\partial z}{\partial x_k} \right\}, \left\{ \frac{\partial^2 z}{\partial x_k \partial x_l} \right\} \right).$$

$\{\dot{g}_{ij}\}$ is represented in (g) by $\{\varphi_\mu\}$, so this shows that $\{\dot{z}_\alpha\}$ depends linearly on $\{\dot{g}_{ij}\}$. The $F_{\alpha ij}$ are analytic functions of the first and second order derivatives of the imbedding functions, so long as these are such that (B3)–(B5) is non-singular.

To recapitulate, (B7) indicates the form and behavior of the solution of the system:

(B8a) $$\sum_\alpha \frac{\partial z_\alpha}{\partial x_i} \dot{z}_\alpha = 0$$

(B8b) $$-2 \sum_\alpha \frac{\partial^2 z_\alpha}{\partial x_i \partial x_j} \dot{z}_\alpha = \dot{g}_{ij}$$

(B8c) $$\sum_\alpha (\dot{z}_\alpha)^2 = \text{minimum, subject to (a) and (b).}$$

The solution is an imbedding perturbation rate $\{\dot{z}_\alpha\}$ which leads to the rate $\{\dot{g}_{ij}\}$ of change of the metric induced by the imbedding. We call this method of determining a perturbation rate $\{\dot{z}_\alpha\}$ the "perturbation device."

Notational Conventions

The work below becomes almost entirely the treatment of a problem in analysis, so a different (condensed) notation is appropriate. We shall drop the coordinate indices generally. Thus

$\{z_\alpha\}$ becomes z, $\{\dot{z}_\alpha\}$ becomes $\dot{z}$, $\left\{ \frac{\partial z_\alpha}{\partial x_i} \right\}$ becomes z', $\left\{ \frac{\partial^2 z_\alpha}{\partial x_i \partial x_j} \right\}$

becomes z'', etc.

We now write (B7) as

(B9) $$\dot{z} = F(z', z'') \boxtimes \dot{g}.$$

$\boxtimes$ indicates the (contracting) tensor product acting between F and $\dot{g}$. (B9) is the solution of

(a) $$z' \circ \dot{z} = 0$$
(B10) (b) $$-2z'' \circ \dot{z} = \dot{g}$$
(c) $$|\dot{z}| = \text{minimum, subject to (a) and (b).}$$

Here ∘ indicates the scalar product, summation over the index α of E^m. We also have

(B11)
$$\dot{g} = 2z' \otimes \dot{z}'$$

as a modified condensation of (B2). Here $\otimes$ is a symmetrizing tensor product accompanied by a summation (as with ∘).

We shall deal with many inequalities on the sizes of functions and derivatives. Generally these will be tied to the parameter θ, which controls the smoothing operator (see Part A). θ will play a dual role, being both the parameter of smoothing and the parameter of the process. Dot will denote $\partial/\partial\theta$. [e.g.: $\dot{z} = \partial z/\partial\theta$]. The process will begin with a specific value of θ, called θ_0, and end with $\theta = \infty$.

Our canonical notation for bounds on the sizes of functions and their derivatives is explained by illustration:

$$T \lesssim K \left[\theta \big|_{2,\,4}^{-1,\,2} \right]$$

indicates a whole system of bounds on the tensor T and its derivatives measured in terms of the $\text{size}^{(r)}$ concept of Part A (think of T as varying with θ), which are

$$\text{size}^{(0)}\ T \lesssim K\theta^{-1}$$
$$\text{size}^{(1)}\ T \lesssim K\theta^{-1}$$
$$\text{size}^{(2)}\ T \lesssim K\theta^{-1}$$
$$\text{size}^{(3)}\ T \lesssim K$$
$$\text{size}^{(4)}\ T \lesssim K\theta$$
$$\text{size}^{(5)}\ T \lesssim K\theta^{2}.$$

In general, if the symbol is $[\theta|_{r,\,s}^{p;\,q}]$, where r and s are integers and $0 \leq r \leq s$, $p \leq q$, the exponent of θ is p for $\text{size}^{(0)}$, $\text{size}^{(1)}$, $\cdots$, $\text{size}^{(r)}$. Then from $\text{size}^{(r)}$ to $\text{size}^{(s)}$ the exponent increases in arithmetic progression in such a way that we have θ^q for $\text{size}^{(s)}$. Almost always this increase will be one unit for each increase of the order of differentiation, as in the illustration. In other words, we usually have $q - p = s - r$, and also q and p are usually integers.

Most often we can abbreviate $[\theta|_{r,\,s}^{p;\,q}]$ to $[_{r,\,s}^{p;\,q}]$ and understand that θ is involved. Also we shorten symbols such as $[_{3,\,3}^{0,\,0}]$ to $[_3^0]$.

The Perturbation Process

As was remarked above, this process uses the perturbation device and the smoothing operator. It also uses "feed-back."

The equations defining the process are listed below:

(B12) $\zeta = S_\theta z$

(B13) $\dot{z} = F(\zeta', \zeta'') \boxtimes M$

$\qquad E = M - \dot{g}$

(B14) $\qquad = 2(\zeta - z)' \otimes \dot{z}'$ (equivalence shown below)

(B15) $u(p) = $ a special C^∞ function, nondecreasing everywhere and monotone increasing for $0 \leqq p \leqq 1$. Also, $u(p) = 0$ for $p \leqq 0$ and $u(p) = 1$ for $p \geqq 1$. To be specific think of $u(p) = \psi(2 - p)$, where ψ is the special function of Part A.

(B16) $$L(\theta) = \int_{\theta_0}^{\theta} E(\bar{\theta}) u(\theta - \bar{\theta}) \, d\bar{\theta}$$

(B17) $G = $ the desired total change of the metric tensor (a symmetric co-variant tensor).

(B18)

(a) $\displaystyle\int_{\theta_0}^{\theta} M(\bar{\theta}) \, d\bar{\theta} = u(\theta - \theta_0) S_\theta G + S_\theta L(\theta)$, or

(b) $M = \dot{u}(\theta - \theta_0) S_\theta G + u(\theta - \theta_0) \dot{S}_\theta G + \dot{S}_\theta L + S_\theta \dot{L}$, or

(c) $M = \dot{u}(\theta - \theta_0) S_\theta G + u(\theta - \theta_0) \dot{S}_\theta G + \dot{S}_\theta L$

$$\qquad\qquad + S_\theta \int_{\theta_0}^{\theta} E(\bar{\theta}) \dot{u}(\theta - \bar{\theta}) \, d\bar{\theta}.$$

Interpretation of the Quantities and Equations

Each of the quantities has its interpretation, but our interpretation may seem unilluminating to many. ζ is a smooth approximation to the imbedding. M is the rate of metric change that is being "attempted." Since the actual rate, $\dot{z}$, of perturbation of the imbedding is computed from a formula, which is like (B9) but has ζ in place of z, we cannot expect that the actual rate, $\dot{g}$, of metric change will be the same as M. If the imbedding function were ζ instead of z then $\dot{g}$ would be M. That is, (B13) gives the correct rate of imbedding perturbation to accomplish the metric change M, provided ζ is the imbedding. This implies that

$$(B19) \qquad\qquad M = 2\zeta' \otimes \dot{z}'.$$

E is the error rate, or the excess of the attempted metric change rate over the actual rate, so $E = M - \dot{g}$. Since we have (B11),

$$\dot{g} = 2z' \otimes \dot{z}',$$

we can subtract this from (B19) to obtain the alternate E formula:

$$E = 2(\zeta - z)' \otimes \dot{z}'.$$

This formula is the more useful one for estimating the size of E.

L represents accumulated error. It is not the total accumulated error, but it includes all the error incurred up to $\theta - 1$. Between $\theta - 1$ and θ it includes only part of the error. Effectively, there is a lag in the inclusion of error in L. This effect is accomplished by the weighting function $u(\theta - \bar{\theta})$ in (B16). It is not really necessary to define L so elaborately. It could be the total error. The advantage of the more elaborate definition is that it ultimately makes it easier to view the process as controlled by a set of very tame integral equations. $\dot{L}$ has the simple integral expression

$$\dot{L} = \int_{\theta_0}^{\theta} E(\bar{\theta})\dot{u}(\theta - \bar{\theta})\,d\bar{\theta}.$$

If L were the total error $\dot{L}$ would be E. Then it would seem that M was defined from E and E from M. We avoid this complication with the definition we use.

The definition of M is based on the principle of "feeding in" the smoother parts of the desired metric correction first, saving the rougher parts for later. Referring to (B18a), we can consider $\int_{\theta_0}^{\theta} M$ the total "attempted" metric change from the beginning of the process at θ_0 to the current situation at θ. This is set equal to $S_\theta L + u(\theta - \theta_0)S_\theta G$. So it is the smooth part of L plus the smooth part of G weighted by the coefficient $u(\theta - \theta_0)$. The reason for attaching $u(\theta - \theta_0)$ to $S_\theta G$ in this formula is the simple one that for $\theta = \theta_0$ both sides of the equation (B18a) should be zero. At the beginning of the process a finite portion of G, specifically $S_{\theta_0} G$, is considered smooth enough to be "fed in." But it must be fed in gradually, so we use $u(\theta - \theta_0)$ to make the process start gradually. For $\theta \geqq \theta_0 + 1$ this factor $u(\theta - \theta_0)$ is just $+1$ and is irrelevant.

We can see how these definitions should work out if the process is convergent. The total metric change accomplished by the process from its start at θ_0 to the limit as $\theta \to \infty$ will be $\int_{\theta_0}^{\infty} \dot{g}\, d\theta$. From (B14),

(B20) $$\dot{g} = M - E,$$

therefore

(B21) $$\int_{\theta_0}^{\infty} \dot{g}\, d\theta = \int_{\theta_0}^{\infty} M - \int_{\theta_0}^{\infty} E$$
$$= u(\infty) S_{\infty} G + S_{\infty} L(\infty) - \int_{\theta_0}^{\infty} E$$
$$= G + L(\infty) - L(\infty)$$
$$= G.$$

To put $L(\infty)$ for $\int_{\theta_0}^{\infty} E$ requires the assumption that $E \to 0$ as $\theta \to \infty$. (B21) verifies the general design of our "feed-back" process, but of course the main task is the proof of convergence. These remarks on interpretation should not be regarded as if presented as proofs. However, we shall use the equivalence of the two formulas of (B14).

To prove that the process works we first derive a set of appropriate *a priori* bounds on the quantities involved which would be satisfied if the equations defining the process have a solution up to some value θ_1 of θ. Second we prove a local continuation theorem about solutions of the equation system. The combination of the bounds and the local continuation theorem gives us the existence and uniqueness of the solution for all values of θ, but we assume that G is sufficiently small and that θ_0 is properly chosen in obtaining this result.

The Estimates

These estimates, or bounds, form a self-interacting system because the size of each function tends to depend on the sizes of the others. We assume the system (B12) through (B18) has a solution for $\theta_0 \leqq \theta \leqq \theta_1$ and assume that the quantities E, M, ζ, z, L, etc., satisfy certain bounds in this range. Then we compute new bounds on the quantities from the defining equations. Finally, we show that when G is sufficiently small and θ_0 properly chosen there exists a set of bounds which is satisfied by the initial values of the quantities and is such that the rederived bounds computed from the defining equations are all smaller than this original set of bounds.

The first bound is

(B22) $$\zeta - z_0 \lesssim \varepsilon \begin{bmatrix} 0 \\ 2 \end{bmatrix}.$$

z_0 stands for $z(\theta_0)$, the initial value of the imbedding function. We assume that z_0 is analytic and such that (B8a, b) is non-singular. $F(z', z'')$ will be well behaved when z, z', z'' are near z_0, z_0', z_0'', respectively. So (B22) is designed to insure the good behavior of $F(\zeta', \zeta'')$. ε must be a sufficiently small constant so that $F(\zeta', \zeta'')$ will be well behaved when (B22) holds.

There will be some value θ_α (assume $\theta_\alpha \geqq 1$) such that for $\theta \geqq \theta_\alpha$ we have $S_\theta z_0 - z_0 \leqq \varepsilon/2 \begin{bmatrix} 0 \\ 2 \end{bmatrix}$. So we make a requirement:

(B23) $$\theta_0 \geqq \theta_\alpha.$$

With this requirement we can satisfy (B22) by keeping $z - z_0$ small enough. Let

(B24) $$z_0 \lesssim \alpha \begin{bmatrix} 0,1 \\ 3,4 \end{bmatrix}.$$

This is for notational convenience only. z_0 is of course independent of θ. The bound α should be chosen so that (B24) holds for all $\theta \geqq 1$. For $z - z_0$ we assume a bound

(B25) $$z - z_0 \lesssim \beta \begin{bmatrix} 0,1 \\ 3,4 \end{bmatrix}.$$

Note that if (B23) holds and β is small enough then (B22) holds. Adding (B24) and (B25),

$$z \lesssim (\alpha + \beta) \begin{bmatrix} 0,1 \\ 3,4 \end{bmatrix}$$

(B26) $$\lesssim \xi \begin{bmatrix} 0,1 \\ 3,4 \end{bmatrix}.$$

ξ is employed for notational advantages.

Other bounds used are:

(B27) $$L \lesssim \lambda \begin{bmatrix} 0 \\ 3 \end{bmatrix}$$

(B28) $$M \lesssim \mu \begin{bmatrix} -4,0 \\ 0,4 \end{bmatrix}$$

(B29)
$$\dot{z} \lesssim \gamma \begin{bmatrix} -4,0 \\ 0,4 \end{bmatrix}$$

(B30)
$$E \lesssim \eta \begin{bmatrix} -5,-2 \\ 0,\ 3 \end{bmatrix}$$

(B31)
$$G \lesssim \delta \begin{bmatrix} 0 \\ 3 \end{bmatrix}.$$

The bound on G is a handle by which we can refer to the size of G, which we shall assume to be as small as necessary to make the process converge. This smallness includes derivatives up to the third order.

The Rederived Bounds

Now we assume that (B22) through (B31) hold for a solution of the equations of the process for $\theta_0 \leqq \theta \leqq \theta_1$ and use the defining equations to compute new "rederived" bounds on the same quantities in terms of the original bounds. The new bounds hold in the same θ range and are distinguished by starred Greek letters.

First consider L. From (B16)

$$L(\theta) \lesssim \int_{\theta_0}^{\theta} \eta \left[\bar{\theta} \Big|_{0,\ 3}^{-5,-2} \right] u(\theta - \bar{\theta})\, d\bar{\theta}.$$

When $k \geqq 2$

$$\int_{\theta_0}^{\theta} (\bar{\theta})^{-k}\, d\bar{\theta} = \frac{\theta_0^{1-k} - \theta^{1-k}}{k-1} \leqq \theta_0^{1-k}.$$

Since $|u| \leqq 1$ and $\theta_0 \geqq 1$ we can put

$$L \lesssim \eta \left[\theta_0 \Big|_{0,\ 3}^{-4,-1} \right]$$

(B32)
$$\lesssim \eta \theta_0^{-1} \begin{bmatrix} 0 \\ 3 \end{bmatrix}$$

$$\lesssim \lambda^* \begin{bmatrix} 0 \\ 3 \end{bmatrix}.$$

To estimate M we use (B18c) and have

$$M \lesssim \dot{u}'(\theta - \theta_0) C_1 \delta \begin{bmatrix} 0,1 \\ 3,4 \end{bmatrix} + C_2 \delta \begin{bmatrix} -4,0 \\ 0,4 \end{bmatrix}$$

(B33)
$$+ C_2 \lambda^* \begin{bmatrix} -4,0 \\ 0,4 \end{bmatrix} + S_\theta \int_{\theta_0}^{\theta} \eta \left[\bar{\theta} \Big|_{0,\ 3}^{-5,-2} \right] \dot{u}'(\theta - \bar{\theta})\, d\bar{\theta}.$$

The first term is obtained from (A15) applied to (B31). The constant C_1 is simply the largest of the coefficients that would come in from (A15). In general, when we have unspecified constants appearing in estimates, we shall simply number them C_1, C_2, $\cdots$ in order of occurrence. We shall not be concerned with the actual sizes of these constants. The second term comes from applying (A16) to (B31), remembering that $|u| \leqq 1$, and the third comes from (A16) and (B32).

For the first term we say

$$\dot{u}'(\theta - \theta_0) C_1 \delta \begin{bmatrix} 0, 1 \\ 3, 4 \end{bmatrix} \lesssim [\theta^4 \dot{u}(\theta - \theta_0)] C_1 \delta \begin{bmatrix} -4, 0 \\ 0, 4 \end{bmatrix}$$

$$\lesssim \max_{\theta_0 \leqq \theta \leqq \theta_0 + 1} \{\theta^4\} \max_\theta \{\dot{u}(\theta - \theta_0)\} C_1 \delta \begin{bmatrix} -4, 0 \\ 0, 4 \end{bmatrix}$$

$$\lesssim C_3 (\theta_0 + 1)^4 \delta \begin{bmatrix} -4, 0 \\ 0, 4 \end{bmatrix}.$$

This brings the first term into the $\begin{bmatrix} -4, 0 \\ 0, 4 \end{bmatrix}$ form, leaving only the fourth term of (B33) to be treated.

We want the fourth term of (B33) majorized by a $\begin{bmatrix} -4, 0 \\ 0, 4 \end{bmatrix}$ term. Let $\theta^* = \max(\theta_0, \theta - 1)$, then since $\dot{u}'(\theta - \bar{\theta}) = 0$ when $\bar{\theta} \leqq \theta - 1$, we can say

$$\text{the fourth term} \lesssim S_\theta \int_{\theta^*}^\theta \eta \left[\bar{\theta}\big|\begin{matrix} -5, -2 \\ 0, \ \ 3 \end{matrix}\right] \dot{u}(\theta - \bar{\theta}) \, d\bar{\theta}$$

$$\lesssim S_\theta \left\{\eta \max_p [\dot{u}(p)] \left[\theta^*\big|\begin{matrix} -5, -2 \\ 0, \ \ 3 \end{matrix}\right]\right\},$$

because the interval of integration is not more than one unit long, i.e., $\theta - \theta^* \leqq 1$. Continuing,

$$\text{the fourth term} \lesssim C_4 \eta \left[\theta^*\big|\begin{matrix} -5, -1 \\ 0, \ \ 4 \end{matrix}\right],$$

where C_4 is to take care of $\max_p \dot{u}(p)$ and the constant coefficients arising from (A15) (which tells us how S_θ acts). Then, since $\theta_0 \leqq \theta^* \leqq \theta$, we can say

$$\text{the fourth term} \lesssim C_4 \eta \theta_0^{-1} (\theta^*/\theta)^{-5} \left[\theta\big|\begin{matrix} -4, 0 \\ 0, 4 \end{matrix}\right]$$

$$\lesssim C_4 \eta \theta_0^{-1} \left(\tfrac{1}{2}\right)^{-5} \left[\theta\big|\begin{matrix} -4, 0 \\ 0, 4 \end{matrix}\right]$$

$$\text{(because } 1 \leqq \theta^* \leqq \theta, \text{ and } \theta^* \geqq \theta - 1\text{)}$$

$$\lesssim C_5 \eta \theta_0^{-1} \begin{bmatrix} -4, 0 \\ 0, 4 \end{bmatrix}.$$

Combining these results we obtain

$$M \lesssim \left\{ C_3(\theta_0 + 1)^4 \delta + C_2 \delta + C_2 \lambda^* + C_5 \theta_0^{-1} \eta \right\} \begin{bmatrix} -4,0 \\ 0,4 \end{bmatrix}$$

(B34)

$$\lesssim \mu^* \begin{bmatrix} -4,0 \\ 0,4 \end{bmatrix}.$$

To estimate $\dot{z}$ we need (B34) and an estimate on ζ. For ζ we write

(B35)
$$\zeta \lesssim C_5 \xi \begin{bmatrix} 0,3 \\ 3,6 \end{bmatrix},$$

by applying (A15) to (B26). The reason for extending this estimate to sixth derivatives is that $\dot{z}$ depends on ζ'', so that the fourth derivatives of $\dot{z}$ depend on the sixth derivatives of ζ.

A derivative of $\dot{z}$ will depend on derivatives of various orders of M and ζ. For example, we can write symbolically

$$\dot{z}' = F(\zeta', \zeta'') \boxtimes M' + (F_{\zeta'}) \zeta'' \boxtimes M + (F_{\zeta''}) \zeta''' \boxtimes M.$$

The function F and its partial derivatives $F_{\zeta'}$ and $F_{\zeta''}$ will be bounded when (B22) holds so we can say

$$\text{size } \dot{z}' \leqq \text{ const. } \mu^* \theta^{-3} + \text{ const. } C_6 \xi \mu^* \theta^{-4} + \text{ const. } C_6 \xi \mu^* \theta^{-4}$$

by applying (B34) and (B35). Notice that the highest (least negative) power of θ comes from the term where M is differentiated. For a higher derivative of $\dot{z}$ there would be many terms and the highest power of θ would appear in the term with the maximum differentiation of M. If we majorize the lower powers of θ by the highest and observe that μ^* would appear once in each term we can put the general estimate for $\dot{z}$ in this form:

$$\dot{z} \lesssim P_1(\xi) \mu^* \begin{bmatrix} -4,0 \\ 0,4 \end{bmatrix}, \quad \text{or}$$

(B36)

$$\lesssim \gamma^* \begin{bmatrix} -4,0 \\ 0,4 \end{bmatrix},$$

where $P_1(\xi)$ is simply some fourth degree polynomial in ξ, the first of a series of such polynomials that we shall use, analogous to the series of numbered constants. Note that $\dot{z}$ has the same θ dependence as M.

We can use (B36) and an estimate

$$\zeta - z \lesssim C_7 \xi \begin{bmatrix} -3,1 \\ 0,4 \end{bmatrix},$$

obtained via (A17) and (B26), to estimate E. Using $E = 2(\zeta - z)' \otimes \dot{z}'$ we obtain

$$E \lesssim 2\left\{C_7\xi\left[\begin{smallmatrix}-3,1\\0,4\end{smallmatrix}\right]\right\}' \otimes \left\{\gamma^*\left[\begin{smallmatrix}-4,0\\0,4\end{smallmatrix}\right]\right\}', \text{ or}$$

$$E \lesssim 2C_7\xi\left[\begin{smallmatrix}-2,1\\0,3\end{smallmatrix}\right] \otimes \gamma^*\left[\begin{smallmatrix}-3,0\\0,3\end{smallmatrix}\right], \text{ or}$$

(B37)
$$E \lesssim C_8\xi\gamma^*\left[\begin{smallmatrix}-5,-2\\0,\ \ 3\end{smallmatrix}\right]$$

$$\lesssim \eta^*\left[\begin{smallmatrix}-5,-2\\0,\ \ 3\end{smallmatrix}\right].$$

This illustrates the pattern of the bound corresponding to the product of two bounds expressed in our notation.

The estimate of z is the most laborious. The simple estimate obtainable through $z = z_0 + \int_{\theta_2}^{\theta} \dot{z}$ is not good enough for the third derivatives of z. We need an estimate on integrals of M as an intermediate step. Here also the direct estimation is inadequate, but we can use (B18a) to say

$$\int_{\theta_2}^{\theta_3} M(\theta)\,d\theta = u(\theta_3 - \theta_0)S_{\theta_3}G + S_{\theta_3}L(\theta_3)$$
$$- u(\theta_2 - \theta_0)S_{\theta_2}G - S_{\theta_2}L(\theta_2).$$

Applying (A15) to the G and L bounds, assuming $\theta_2 \leqq \theta_3$, and using $|u| \leqq 1$, we can say

$$\int_{\theta_2}^{\theta_3} M(\theta)\,d\theta \lesssim C_9(\delta + \lambda^*)\left[\theta_3\Big|_{3,4}^{0,1}\right].$$

The straightforward integration of the M estimate, (B34), yields

$$\int_{\theta_2}^{\theta_3} M(\theta)\,d\theta \lesssim \int_{\theta_2}^{\theta_3} \mu^*\left[\begin{smallmatrix}-4,0\\0,4\end{smallmatrix}\right] d\theta$$
$$\lesssim C_{10}\,\mu^*\left[\theta_2\Big|_{0,\ \ 2}^{-3,-1}\right].$$

This estimate is stated for differentiation only up to the second order because for the third order $\int \theta^{-1}\,d\theta$ would lead to a logarithmic term; and for the fourth order θ_3^{+1} would majorize θ_2^{+1} rather than the θ_2 term majorizing the θ_3 term (which occurs up through the second order derivatives).

Now we have two estimates, one good for lower derivatives, and one good for higher ones. If we add the two estimates we can safely extend the range of the second and modify the first in the range covered by the second. This gives us

$$\text{(B38)} \quad \int_{\theta_2}^{\theta_3} M(\theta)\, d\theta \lesssim C_{11}(\mu^* + \delta + \lambda^*) \left\{ \left[\theta_2 \Big|_{0,4}^{-3,1}\right] + \left[\theta_3 \Big|_{0,4}^{-3,1}\right] \right\}.$$

We also need an $\dot{F}$ estimate in obtaining our z estimate. Symbolically written,

$$\dot{F} = F_{\zeta'}\dot{\zeta}' + F_{\zeta''}\dot{\zeta}''.$$

To use this we must estimate $\dot{\zeta}$, which is $(S_\theta z) = \dot{S}_\theta z + S_\theta \dot{z}$. So from (A16) and (B26) and from (A15) and (B29) we derive

$$\dot{\zeta} \lesssim C_{12}\xi \left[_{0,6}^{-4,2}\right] + C_{13}\gamma \left[_{0,6}^{-4,2}\right]$$
$$\lesssim C_{14}(\xi + \gamma)\left[_{0,6}^{-4,2}\right].$$

Using this $\dot{\zeta}$ estimate we can now estimate $\dot{F}$ in a manner exactly analogous to the manner in which we estimated $\dot{z} = F \boxtimes M$. The result has the form

$$\text{(B39)} \quad \begin{aligned} \dot{F} &\lesssim P_2(\xi)(\xi + \gamma)\left[_{0,5}^{-3,2}\right] + P_3(\xi)(\xi + \gamma)\left[_{0,4}^{-2,2}\right] \\ &\lesssim P_4(\xi)(\xi + \gamma)\left[_{0,4}^{-2,2}\right]. \end{aligned}$$

Finally, we need an estimate on F itself. Here the highest power of θ will come from the maximum differentiation of ζ''. So the estimate takes the form

$$\text{(B40)} \quad F \lesssim P_5(\xi)\left[_{1,4}^{0,3}\right]$$

and has the same θ dependence as ζ''.

The z Estimate

We actually estimate $z - z_0$, then z is easily estimated from this result. First,

$$\begin{aligned} z(\theta_3) - z_0 &= \int_{\theta_0}^{\theta_3} \dot{z}\, d\theta \\ &= \int_{\theta_0}^{\theta_3} F \boxtimes M\, d\theta \\ &= \int_{\theta_0}^{\theta_3} F \boxtimes \left[-\frac{\partial}{\partial \theta_2} \int_{\theta_2}^{\theta_3} M\, d\theta \right] d\theta \end{aligned}$$

$$= -\int_{\theta_0}^{\theta_3} F \boxtimes \left(\frac{\partial}{\partial \theta_2} \int_{\theta_2}^{\theta_3} M \, d\theta \right) d\theta.$$

Now we apply integration by parts and have

$$z(\theta_3) - z_0 = -\left[F \boxtimes \int_{\theta_2}^{\theta_3} M \, d\theta \right]_{\theta_2=\theta_0}^{\theta_2=\theta_3} + \int_{\theta_0}^{\theta_3} \dot{F} \boxtimes \left(\int_{\theta_2}^{\theta_3} M \, d\theta \right) d\theta_2.$$

The first term can be evaluated and we obtain

180

(B41)

$$z(\theta_3) - z_0 = F(\zeta'(\theta_0), \zeta''(\theta_0)) \boxtimes \int_{\theta_0}^{\theta_3} M \, d\theta$$

$$+ \int_{\theta_0}^{\theta_3} \left\{ \dot{F}(\text{at } \theta_2) \boxtimes \int_{\theta_2}^{\theta_3} M \, d\theta \right\} d\theta_2.$$

At this point we insert the estimates (B40) for F, (B39) for $\dot{F}$, and (B38) for $\int M \, d\theta$. This yields

(B42)

$$z - z_0 \lesssim P_5(\xi) \left[\theta_0 \big|_{1,4}^{0,3} \right]$$

$$\boxtimes C_{11}(\mu^* + \delta + \lambda^*) \left\{ \theta_0 \left[{}_{0,4}^{-3,1} \right] + \left[\theta_3 \big|_{0,4}^{-3,1} \right] \right\}$$

$$+ \int_{\theta_0}^{\theta_3} \left\{ P_4(\xi)(\xi + \lambda) \left[\theta_2 \big|_{0,4}^{-2,2} \right] \right\}$$

$$\boxtimes C_{11}(\mu^* + \delta + \lambda^*) \left\{ \left[\theta_2 \big|_{0,4}^{-3,1} \right] + \left[\theta_3 \big|_{0,4}^{-3,1} \right] \right\} d\theta_2.$$

Call the two terms on the right T_1 and T_2. Using $\theta_3 \geqq \theta_0 \geqq 1$ we can weaken and simplify T_1 and have

$$T_1 \lesssim P_5(\xi)(\mu^* + \delta + \lambda^*) \left[\theta_3 \big|_{3,4}^{0,1} \right].$$

T_2 must be handled with more care. Each derivative of T_2 would correspond to the sum of several integrals involving various powers of θ_2 and θ_3 in the integrand. For the rth order derivatives of T_2 there are terms of the form

$$\int_{\theta_0}^{\theta_3} \text{constant} \cdot \theta_2^s \cdot \theta_3^{r-s-5} \, d\theta_2,$$

where $r = 0, 1, 2, 3, 4$ and either $s = r - 5$ (which makes $r - s - 5 = 0$) or s satisfies $-2 \leqq s \leqq r-2$. These two alternatives correspond to the two expressions $\left[\theta_2 \big|_{0,4}^{-3,1} \right]$ and $\left[\theta_3 \big|_{0,4}^{-3,1} \right]$ which are added in T_2. These integrals give varied terms and we can handle the situation most clearly by simply listing all cases. (This is done in Figure 1.)

	$r=0$	$r=1$	$r=2$	$r=3$	$r=4$
$s=r-5$	$\frac{1}{4}\theta_0^{-4}$	$\frac{1}{3}\theta_0^{-3}$	$\frac{1}{2}\theta_0^{-2}$	θ_0^{-1}	$\log(\theta_3/\theta_0)$
$s=\ -2$	$\theta_0^{-1}\theta_3^{-3}$	$\theta_0^{-1}\theta_3^{-2}$	$\theta_0^{-1}\theta_3^{-1}$	θ_0^{-1}	$\theta_0^{-1}\theta_3$
-1		$\log(\theta_3/\theta_0)\theta_3^{-3}$	$\log(\theta_3/\theta_0)\theta_3^{-2}$	$\log(\theta_3/\theta_0)\theta_3^{-1}$	$\log(\theta_3/\theta_0)$
0			θ_3^{-2}	θ_3^{-1}	1
1				$\frac{1}{2}\theta_3^{-1}$	$\frac{1}{2}$
2					$\frac{1}{3}$
majorizer	θ_0^{-4}	θ_0^{-3}	θ_0^{-2}	θ_0^{-1}	$\theta_0^{-1}\theta_3$

Figure 1.

By using the majorizing terms listed at the bottom of the chart we can say

$$T_2 \lesssim P_7(\xi)(\xi+\gamma)(\mu^*+\delta+\lambda^*)\theta_0^{-1}\left[\theta_3\big|_{3,4}^{0,1}\right].$$

Because the pattern of powers of θ_0 does not fit into our notation scheme we have simply used the highest power (θ_0^{-1}) which occurs.

Now if we add the T_1 and T_2 estimates we get an estimate for $z(\theta_3)-z_0$:

$$
\text{(B43)} \qquad
\begin{aligned}
z(\theta_3)-z_0 &\lesssim P_8(\xi)(1+\xi+\gamma)(\mu^*+\delta+\lambda^*)\left[\theta_3\big|_{3,4}^{0,1}\right], \quad \text{or}\\
z-z_0 &\lesssim \beta^*\left[\begin{smallmatrix}0,1\\3,4\end{smallmatrix}\right].
\end{aligned}
$$

Since $z=z_0+(z-z_0)$ we can say

$$
\text{(B44)} \qquad
\begin{aligned}
z &\lesssim (\alpha+\beta^*)\left[\begin{smallmatrix}0,1\\3,4\end{smallmatrix}\right]\\
&\lesssim \xi^*\left[\begin{smallmatrix}0,1\\3,4\end{smallmatrix}\right].
\end{aligned}
$$

To conclude the rederivation of bounds we must consider the requirement (B22). We have stipulated $\theta_0 \geqq \theta_a$ so that

$$S_\theta z_0 - z_0 \leqq \varepsilon/2\left[\begin{smallmatrix}0\\2\end{smallmatrix}\right].$$

Applying (A15) to (B43) we see that

$$\zeta - S_\theta z_0 = S_\theta(z - z_0) \leqq C_{15}\beta^* \begin{bmatrix} -3,1 \\ 0,4 \end{bmatrix}.$$

Adding inequalities,

(B45)
$$\zeta - z_0 = (\zeta - S_\theta z_0) + (S_\theta z_0 - z_0) \leqq (C_{15}\beta^* + \varepsilon/2) \begin{bmatrix} 0 \\ 2 \end{bmatrix}$$
$$\leqq \varepsilon^* \begin{bmatrix} 0 \\ 2 \end{bmatrix};$$

We hope, of course, that $\varepsilon^* \leqq \varepsilon$.

Strong Consistency of the Bounds

We verify here that if δ is sufficiently small we can choose θ_0 so that $\theta_0 \geqq \theta_a$ as required and so that all the rederived bounds are smaller than the originals: $\lambda^* < \lambda$, $\mu^* < \mu$, $\varepsilon^* < \varepsilon$, etc. For clarity the formulas for all the rederived bounds are assembled here:

$$\lambda^* = \eta \theta_0^{-1} \tag{B32}$$

$$\mu^* = C_3(\theta_0 + 1)^4 \delta + C_2 \delta + C_2 \lambda^* + C_5 \theta_0^{-1} \eta \tag{B34}$$

$$\gamma^* = P_1(\xi)\mu^* \tag{B36}$$

(B46)
$$\eta^* = C_8 \xi \gamma^* \tag{B37}$$

$$\beta^* = P_8(\xi)(1 + \xi + \gamma)(\mu^* + \delta + \lambda^*) \tag{B43}$$

$$\xi^* = \alpha + \beta^* \quad (\xi = \alpha + \beta) \tag{B44}$$

$$\varepsilon^* = \varepsilon/2 + C_{15}\beta^* \tag{B45}.$$

Consider any set $\lambda, \mu, \gamma, \eta, \beta, \xi, \varepsilon$ (with $\xi > \alpha$) of positive original bounds and consider the behavior of the rederived bounds as $\theta_0 \to \infty$ and as $\delta \to 0$ for each θ_0 value. That is, consider

$$\lim_{\theta_0 \to \infty} \lim_{\delta \to 0}$$

of each rederived bound.

First we see that

$$\lim_{\theta_0 \to \infty} \left[\lim_{\delta \to 0} \lambda^* \right] = 0,$$

or $\lambda^* \to 0$. Since $\lambda^* \to 0$ we can see that $\mu^* \to 0$. Therefore $\gamma^* \to 0$, whence $\eta^* \to 0$. Because $\mu^*, \delta, \lambda^* \to 0$ we see that $\beta^* \to 0$. Therefore $\xi^* \to \alpha$ and $\varepsilon^* \to \varepsilon/2$. These observations make it clear that, regardless of the sizes of the bounds $\lambda, \mu, \cdots$ assumed originally, if

θ_0 is taken large enough and δ is sufficiently small the rederived bounds $\lambda^*, \mu^*, \cdots$ will all be smaller than the original bounds.

Existence of the Solution

To use the results above we need a general understanding of the equation system defining our perturbation process. We must know there will be a local continuation whenever we have a well-behaved solution of the system in an interval $\theta_0 \leqq \theta < \theta_1$. Actually, the equation system is a very innocuous one. The smoothing makes the solutions analytic functions and makes them very well behaved, at least over small ranges of θ.

The form which the system takes above, as (B12) through (B18), does not reveal its true character very directly. The extent of the taming effect of S_θ is not fully apparent. Another formulation, given below, makes this tameness apparent.

A removable aspect of the system as presented above is the differentiation with respect to spatial coordinates, which is indicated by priming, as with ζ' or z'. This differentiation makes the system look like a partial differential equation system, a type of system where local continuations of solutions often do not generally exist.

If a quantity is defined by smoothing, such as $\zeta = S_\theta z$, then its space derivative can be expressed as the result of an appropriate operator's action on the original quantity. So let us say

$$\zeta' = S_\theta' z.$$

For higher derivatives we have analogous operators S_θ'', S_θ''', etc.; and for $\dot{S}_\theta$ there is a similar series $\dot{S}_\theta'$, $\dot{S}_\theta''$, $\cdots$. These derived operators also have smoothing properties. By using them we can recast the system as an innocuous functional-integral-equation system.

Regard z, z', L and $\dot{L}$ as basic quantities. Then the others are expressible as functions or functionals of these four, either direct or indirect:

$$\zeta = S_\theta z, \quad \zeta' = S_\theta' z, \quad \zeta'' = S_\theta'' z, \quad \zeta''' = S_\theta''' z$$
$$M = S_\theta \left(\dot{u}(\theta - \theta_0)G + \dot{L} \right) + \dot{S}_\theta (u(\theta - \theta_0)G + L)$$
$$M' = S_\theta' \left(\dot{u}(\theta - \theta_0)G + \dot{L} \right) + \dot{S}_\theta' (u(\theta - \theta_0)G + L)$$
$$\dot{z} = F(\zeta', \zeta'') \boxtimes M$$

$$\dot{z}' = (F_{\zeta'} \cdot \zeta'' + F_{\zeta''} \cdot \zeta''') \boxtimes M + F \boxtimes M'$$
$$\dot{g} = 2\dot{z}' \otimes \dot{z}'$$
$$E = M - \dot{g} \quad \text{or} \quad E = 2(\zeta' - z') \otimes \dot{z}'.$$

The basic quantities are to be equal to certain integrals:

$$z = z_0 + \int_{\theta_0}^{\theta} \dot{z}, \quad z' = z_0' + \int_{\theta_0}^{\theta} \dot{z}'.$$

$$L = \int_{\theta_0}^{\theta} u(\theta - \bar{\theta}) E(\bar{\theta}) \, d(\bar{\theta}), \quad \dot{L} = \int_{\theta_0}^{\theta} \dot{u}(\theta - \bar{\theta}) E(\bar{\theta}) \, d\bar{\theta}.$$

Now, since all the functions, functionals, and operations involved are well-behaved, in fact, analytic, we have a well-behaved integral equation system. Indeed, the system is especially tame. Since $\dot{z}$ and $\dot{z}'$ are defined through smoothed quantities exclusively and since z_0 and z_0' are analytic, therefore z and z' are smooth and analytic (as functions of the space variables). Consequently $\dot{g}$ and E are smooth. So L and $\dot{L}$ are also smooth. Thus all the quantities are kept smooth by the presence of S_θ in the equations.

Continuation

Suppose the system has a solution for $\theta_0 \leqq \theta < \theta_1$. We can show that this solution will satisfy the bounds λ, μ, etc. The bounds will certainly hold at the beginning of the solution, so they will hold in an interval $\theta_0 \leqq \theta \leqq \varphi$ where $\theta_0 < \varphi < \theta_1$, if they do not hold up to θ_1. But now the smaller bounds λ^*, μ^*, $\cdots \varepsilon^*$ must hold in $\theta_0 \leqq \theta \leqq \varphi$. Therefore the larger bounds (from continuity) must hold in some interval past φ. This contradiction proves that the larger bounds, and therefore also the smaller rederived bounds, hold in the whole range $\theta_0 \leqq \theta < \theta_1$ of the solution.

Now suppose θ_1 is the limit of continuation of the solution. There will certainly be a solution in the closed interval $\theta_0 \leqq \theta \leqq \theta_1$ because of the bounds and the tameness of the system. But a standard argument, such as the Picard method or the functional fixed point approach, will show that the system has a local continuation past θ_1. Furthermore, the continuation of a solution will be unique. Therefore we have contradicted the hypothesis and we see that the solution will exist, will be unique, and will satisfy the bounds for all values $\theta \geqq \theta_0$.

Convergence to Isometry

We must show that the imbeddings $z(\theta)$ occurring in the perturbation process tend to a limit imbedding which realizes the desired metric $G + g_0$, where g_0 is the metric of the initial imbedding.

The Cauchy criterion approach to the proof of convergence of the imbedding requires us to consider

$$z(\theta_2) - z(\theta_1), \quad \text{which} \quad = \int_{\theta_1}^{\theta_2} \dot{z}(\theta)\, d\theta,$$

which

$$\lesssim \int_{\theta_1}^{\theta_2} \gamma \begin{bmatrix} -4,0 \\ 0,4 \end{bmatrix} d\theta,$$

or

$$\lesssim \gamma \left[\theta_1\right|_{0,\ 2}^{-3,-1}\right].$$

This estimate on $z(\theta_2) - z(\theta_1)$ and its derivatives is good enough to show that the imbedding $z(\theta)$ and its first and second derivatives converge to a limit $z(\infty)$. The integration of the $\dot{z}$ bound is too crude an approach for the third derivatives.

To check the convergence of the metric, $g(\theta)$, induced by the imbedding to the desired limit, $g_0 + G$, observe that

$$g(\theta_1) = g_0 \int_{\theta_0}^{\theta_1} \dot{g}\, d\theta$$

(B48)
$$= g_0 + \int_{\theta_0}^{\theta_1} (M - E)\, d\theta$$

$$= g_0 + \int_{\theta_0}^{\theta_1} M\, d\theta - \int_{\theta_0}^{\theta_1} E\, d\theta.$$

Now apply (B18a):

$$g(\theta_1) = g_0 + u(\theta_1 - \theta_0) S_{\theta_1} G + S_{\theta_1} L(\theta_1) - \int_{\theta_0}^{\theta_1} E\, d\theta.$$

Assume $\theta_1 \geqq \theta_0 + 1$ so that $u(\theta_1 - \theta_0) = 1$. Then we can say

$$g(\theta_1) = g_0 + S_{\theta_1} G + S_{\theta_1} L(\theta_1) - \int_{\theta_0}^{\theta_1} [1 - u(\theta_1 - \theta)] E(\theta)\, d\theta$$

$$- \int_{\theta_0}^{\theta_1} u(\theta_1 - \theta) E(\theta)\, d\theta.$$

The first integrand vanishes for $\theta \le \theta_1 - 1$, and the second integral is $L(\theta_1)$, so

$$g(\theta_1) = g_0 + G + (S_{\theta_1} - 1)[G + L(\theta_1)]$$

186

(B49)
$$+ \int_{\theta_1 - 1}^{\theta_1} [u(\theta_1 - \theta) - 1] E(\theta)\, d\theta.$$

Let us call the last two terms remainders R_1 and R_2. Then by (A17) and (B27) and (B31) we see that

$$R_1 \lesssim C_{16}(\delta + \lambda)\left[\theta_1\Big|_{\ 0,3}^{-3,0}\right].$$

From (B30),

$$R_2 \lesssim \eta\left[\theta_1 - 1\Big|_{\ 0,\ 3}^{-5,-2}\right].$$

The bound on R_2 is good enough for third derivatives but the R_1 bound is not. So we see that the metric $g(\theta)$ converges to the desired limit metric and also see that the first and second derivatives converge to the corresponding derivatives of the limit metric. But we do not yet know that the third derivatives converge. The limit metric, $g_0 + G$, is C^3, of course.

More Refined Results

This section of Part B can be passed over without losing the continuity of the paper. Here we show that the limit imbedding is actually always C^3 and we treat the cases where G is $C^4, C^5, \cdots, C^\infty$.

The cases where G is C^k are treated by an inductive method. The C^k case is handled with the aid of the results of the C^{k-1} case. We can illustrate all the essential features of the induction (with minimum notational difficulty) by considering the step from C^3 to C^4. To begin, we can assume bounds

(B50)
$$G \lesssim \delta'\begin{bmatrix} 0 \\ 4 \end{bmatrix}$$

$$z_0 \lesssim \alpha'\begin{bmatrix} 0 \\ 4 \end{bmatrix}$$

that have the same form as (B24) and (B31), with 4 replacing 3. However, we make a point of not requiring δ' to be small.

Assume that the perturbation process is applied just as if G were only C^3. We will have the bounds (B22) through (B31) on the quantities involved. Our result will be that we can deduce a new set of bounds that are analogous to these except for the systematic replacement of 3 by 4, 4 by 5, etc., in the indices referring to order of differentiation.

In the above work which computed the set of rederived bounds λ^*, μ^*, etc., wherever an estimate was derived for a quantity defined by smoothing, the estimate could have been derived for any particular upper limit on the order of differentiation. Instead of (B34) we could have derived

$$M \lesssim \bar{\mu} \begin{bmatrix} -4,1 \\ 0,5 \end{bmatrix}.$$

This is not the kind of estimate we ultimately want for M. That would be an $\begin{bmatrix} -5,0 \\ 0.5 \end{bmatrix}$ estimate. This is just a useful intermediate step. Similarly we can say

$$\zeta \lesssim \bar{C}_6 \xi \begin{bmatrix} 0,4 \\ 3,7 \end{bmatrix}$$

and

$$\dot{z} \lesssim \bar{\gamma} \begin{bmatrix} -4,2 \\ 0,6 \end{bmatrix}.$$

By integrating this and using (B50) to estimate fifth derivatives and by using (B26) for the lower derivatives we get a crude z bound:

$$z \lesssim \xi \begin{bmatrix} 0,2 \\ 3,5 \end{bmatrix}.$$

From this we have

$$\zeta - z \lesssim \bar{C}_7 \xi \begin{bmatrix} -3,2 \\ 0,5 \end{bmatrix}.$$

Since $E = 2(\zeta - z)' \otimes \dot{z}'$, it can be estimated from our above results and we find

$$E \lesssim \bar{\eta} \begin{bmatrix} -5,-1 \\ 0, \ \ 4 \end{bmatrix}.$$

A weaker E estimate, more conveniently manipulated, is

$$E \lesssim \bar{\eta} \begin{bmatrix} -4\frac{1}{2},-\frac{1}{2} \\ 0, \ \ 4 \end{bmatrix}.$$

This weakened estimate behaves nicely in an integrand and we get new L and $\dot{L}$ estimates via the L and $\dot{L}$ formulas. These are

(B51)
$$L \lesssim \bar{\lambda} \begin{bmatrix} 0, \frac{1}{2} \\ 3, 4 \end{bmatrix}$$

$$\dot{L} \lesssim \bar{C}_5 \bar{\eta} \begin{bmatrix} -4\frac{1}{2}, -\frac{1}{2} \\ 0, \quad 4 \end{bmatrix}.$$

Now reconsider the estimation of M and use the new L and $\dot{L}$ bounds and the new G bound, (B50). One gets

$$M \lesssim \mu^{\#} \begin{bmatrix} -4\frac{1}{2}, \frac{1}{2} \\ 0, \quad 5 \end{bmatrix}.$$

From this we obtain

$$\dot{z} \lesssim \gamma^{\#} \begin{bmatrix} -4\frac{1}{2}, \frac{1}{2} \\ 0, \quad 5 \end{bmatrix},$$

then

$$E \lesssim \eta^{\#} \begin{bmatrix} -5\frac{1}{2}, -1\frac{1}{2} \\ 0, \quad 4 \end{bmatrix}.$$

This sharper E bound yields improved bounds for L and $\dot{L}$:

$$L \lesssim \lambda' \begin{bmatrix} 0 \\ 4 \end{bmatrix}$$

$$\dot{L} \lesssim \bar{C}_5 \eta^{\#} \begin{bmatrix} -5\frac{1}{2}, -1\frac{1}{2} \\ 0, \quad 4 \end{bmatrix}.$$

Now from these and the new G bound we can derive

$$M \lesssim \mu' \begin{bmatrix} -5, 0 \\ 0, 5 \end{bmatrix},$$

then

$$\dot{z} \lesssim \gamma' \begin{bmatrix} -5, 0 \\ 0, 5 \end{bmatrix},$$

then

$$E \lesssim \eta' \begin{bmatrix} -6, -2 \\ 0, \quad 4 \end{bmatrix}.$$

The z estimate depends on estimating $z - z_0$. This can be done exactly like it was done before in computing the rederived bounds. We can use a weak $\dot{\zeta}$ estimate, improved only by extension to sixth derivatives. This gives us a weak $\dot{F}$ estimate, which is however adequate because it is used in combination with an improved estimate on $\int_{\theta_2}^{\theta_3} M \, d\theta$. This improved estimate comes from the strong M, L, and G bounds above (with μ', λ', δ'). The result is of the form

$$z - z_0 \lesssim \beta' \begin{bmatrix} 0,1 \\ 4,5 \end{bmatrix}$$
$$z = \lesssim \xi' \begin{bmatrix} 0,1 \\ 4,5 \end{bmatrix},$$

and this completes the set of new C^4 type estimates (indicated by primed greek letters).

Our result concerning the inductive extension of appropriate bounds to the cases where G is C^k also gives us a result for the C^∞ case. The (typical) induction step from the C^3 case to the C^4 case shown above, involved the use of a new G bound (B50). But we did not assume δ' was small. So our results on the C^k cases with $k > 3$ are really sharper than the exactly analogous results would be, because only G and its derivatives up to the third order need be small.

If G is C^∞ the result of each C^k case is valid, so the imbedding is C^∞.

We can show that when G is C^k the kth derivatives of the imbedding converge so the limit imbedding is C^k. Also the kth derivatives of the metric converge. The argument is the same for any value of k, and since we have not shown this for $k = 3$, we treat that case.

The basic fact is that the limits of the third derivatives of $S_\theta G$ as $\theta \to \infty$ are the third derivatives of G, and that this convergence is uniform. This follows from the uniform continuity of the third derivatives of G, a consequence of compactness. The fact can be symbolized by saying

(B52) $$G - S_\theta G \lesssim \Delta_\theta \begin{bmatrix} 0 \\ 3 \end{bmatrix},$$

where Δ_θ is a constant for each θ value and $\Delta_\theta \to 0$ as $\theta \to \infty$. Nothing can be said about the rapidity of the decrease of Δ_θ; the fact is purely qualitative.

The estimate of (B51) on L, although derived after we assumed G to be C^4, did not depend on this assumption. So we can use it here, where G is again only assumed to be C^3. By applying (A17) to it one can obtain

$$(1 - S_\theta)L(\bar\theta) \lesssim C_{17}\bar\lambda(\bar\theta)^{\frac{1}{2}} \left[\theta_1 \Big|_{0,4}^{-4,0} \right].$$

Assuming $\theta_2 \geqq \theta_1$, it follows that

(B53) $S_{\theta_2} L(\bar{\theta}) - S_{\theta_1} L(\bar{\theta}) \lesssim 2C_{17} \bar{\lambda}(\bar{\theta})^{\frac{1}{2}} \left[\theta_1 \big|_{0,4}^{-4,0} \right].$

We shall need a bound on $L(\theta_2) - L(\theta_1)$. Assume $\theta_2 \geqq \theta_1 \leqq \theta_0 + 1$,

then

$$L(\theta_2) - L(\theta_1) = \int_{\theta_0}^{\theta_2} u(\theta_2 - \theta) E(\theta) \, d\theta - \int_{\theta_0}^{\theta_1} u(\theta_1 - \theta) E(\theta) \, d\theta$$

$$= \int_{\theta_1-1}^{\theta_2} \{u(\theta_2 - \theta) - u(\theta_1 - \theta)\} E(\theta) \, d\theta,$$

because $u(\theta_2 - \theta) = u(\theta_1 - \theta) = 1$ for $\theta \leqq \theta_1 - 1$ and because $u(\theta_1 - \theta) = 0$ for $\theta \geqq \theta_1$. Using (B30),

$$L(\theta_2) - L(\theta_1) \lesssim \int_{\theta_1-1}^{\theta_2} \eta \left[\begin{smallmatrix} -5,-2 \\ 0,\ 3 \end{smallmatrix} \right] d\theta$$

(B54)
$$\lesssim \eta \left[\theta_1 - 1 \Big|_{0,\ 3}^{-4,-1} \right]$$

$$\lesssim 16\eta \left[\theta_1 \Big|_{0,\ 3}^{-4,-1} \right].$$

We must have a more refined estimate on $\int_{\theta_1}^{\theta_2} M$. By (B18a), assuming $\theta_2 \geqq \theta_1 \geqq \theta_0 + 1$,

$$\int_{\theta_1}^{\theta_2} M \, d\theta = S_{\theta_2} G + S_{\theta_2} L(\theta_2) - S_{\theta_1} G - S_{\theta_1} L(\theta_1)$$

(B55)
$$= S_{\theta_2} G - S_{\theta_1} G + S_{\theta_1} [L(\theta_2) - L(\theta_1)]$$
$$+ (S_{\theta_2} - S_{\theta_1}) L(\theta_2)$$
$$= T_a + T_b + T_c.$$

Our principal concern is the third derivatives, which our previous estimate, (B38), merely showed bounded, not decreasing. By (B52),

$$T_a \lesssim (\Delta_{\theta_1} + \Delta_{\theta_2}) \left[\begin{smallmatrix} 0 \\ 3 \end{smallmatrix} \right].$$

If $\theta_1, \theta_2 \to \infty$ the third derivatives of T_a will approach zero, which is what we want.

(B54) is strong enough to show that the third derivatives of T_b will approach zero as $\theta_1, \theta_2 \to \infty$. But T_c requires a more elaborate treatment. We can write

$$T_c = (S_{\theta_2} - S_{\theta_1})L(\theta_2) = (S_{\theta_2} - S_{\theta_1})\left[L(\bar{\theta}) + \{L(\theta_2) - L(\bar{\theta})\}\right]$$

where $\bar{\theta}$ should be regarded as smaller than θ_1. Then by (A15) applied to (B54) and by (B53) we obtain

$$T_c \lesssim C_{18}\eta\left[\bar{\theta}\Big|_{0,3}^{-4,1}\right] + 2C_{17}\bar{\lambda}(\bar{\theta})^{\frac{1}{2}}\left[\theta_1\Big|_{0,4}^{-4,0}\right].$$

Observe now that T_c and its derivatives up to order 3 can be forced to be arbitrarily small if one first chooses $\bar{\theta}$ large enough to make the first term above small and then chooses θ_1 enough larger than $\bar{\theta}$ so that the second term is small. Effectively then, as $\theta_1, \theta_2 \to \infty$ the term T_c and its derivatives up to the third order approach zero.

Since T_a''', T_b''', and T_c''' all approach zero as $\theta_1, \theta_2 \to \infty$, we have shown that

$$\left[\int_{\theta_1}^{\theta_2} M\, d\theta\right]''' \to 0 \quad \text{as} \quad \theta_1, \theta_2 \to \infty.$$

For the lower derivatives of this integral the simple estimate obtained by integrating (B28) is adequate. We can combine the two approaches in a single estimate:

(B56) $$\int_{\theta_1}^{\theta_2} M\, d\theta \lesssim \mu\left[\theta_1\Big|_{0,\ 2}^{-3,-1}\right] + \mu_{\theta_1}\left[\theta_1\Big|_{0,3}^{-3,0}\right].$$

μ_θ is to be a constant for each θ which approaches zero as $\theta \to \infty$.

We show that the third derivatives of the imbedding $z(\theta)$ converge by a Cauchy criterion argument. The difference $z(\theta_3) - z(\theta_1)$ can be expressed in a form exactly analogous to (B41), where θ_1 appears in place of θ_0:

$$z(\theta_3) - z(\theta_1) = F(\zeta'(\theta_1), \zeta''(\theta_1)) \boxtimes \int_{\theta_1}^{\theta_3} M\, d\theta \qquad (= T_1)$$

$$+ \int_{\theta_1}^{\theta_3}\left\{\dot{F}(\text{at } \theta_2) \boxtimes \int_{\theta_2}^{\theta_3} M\, d\theta\right\} d\theta_2 \qquad (= T_2).$$

The analogue of the estimate we obtained, below (B42), for T_1 is not adequate for our needs here. But the analogous T_2 estimate

$$T_2 \leqq P_7(\xi)(\xi + \gamma)(\mu^* + \delta + \lambda^*)\theta_1^{-1}\left[\theta_3\Big|_{3,4}^{0,1}\right]$$

is quite good because θ_1^{-1} becomes small as $\theta_1, \theta_2 \to \infty$.

The estimate (B56) was obtained so that we could handle T_1. We use this with (B40) and F and obtain

$$T_1 \lesssim P_9(\xi)\mu \left[\theta_1\Big|_{0,\ 2}^{-3,-1}\right] + P_{10}(\xi)\mu \left[\theta_1\Big|_{0,\ 3}^{-4,-1}\right]$$

$$+ P_{11}(\xi)\mu_{\theta_1} \left[\theta_1\Big|_{0,3}^{-3,0}\right].$$

This shows that $T_1''' \to 0$ as $\theta_1, \theta_2 \to \infty$ and that is what we need to show that

$$[z(\theta_3) - z(\theta_1)]''' \to 0 \quad \text{as} \quad \theta_1, \theta_2 \to \infty.$$

So we have verified that the third derivatives of the imbedding converge uniformly to the third derivatives of the limit imbedding (which therefore must exist and be continuous).

To see that the third derivatives of the metric converge consider

$$g(\theta_2) - g(\theta_1) = \int_{\theta_1}^{\theta_2} \dot{g}\, d\theta = \int_{\theta_1}^{\theta_2} (M - E)\, d\theta$$

$$= \int_{\theta_1}^{\theta_2} M\, d\theta - \int_{\theta_1}^{\theta_2} E\, d\theta$$

$$\lesssim \mu \left[\theta_1\Big|_{0,\ 2}^{-3,-1}\right] + \mu_{\theta_1} \left\{\theta_1\Big|_{0,3}^{-3,0}\right\} + \eta \left[\theta_1\Big|_{0,\ 3}^{-4,-1}\right],$$

from (B56) and (B30). Since $\mu_{\theta_1} \to 0$ as $\theta_1 \to \infty$, this gives us the convergence of the metric and its derivatives up to the third order.

Summary of Results

The main results of this part of the paper can be summarized in one theorem:

THEOREM I:

Hypotheses:

(1) $\mathfrak{M}$ *is a compact manifold analytically imbedded in a euclidean space.*
(2) *The system* (B8a, b) *of linear equations in the* $\dot{z}_\alpha$ *is non-singular at all points of the imbedding.*
(3) G *is a symmetric covariant tensor on* $\mathfrak{M}$ *representing the change we want to make in the metric induced by the imbedding of* $\mathfrak{M}$. *We want to accomplish this change by modifying the imbedding.*

(4) G is C^k where $3 \leqq k \leqq \infty$.

(5) θ_0 is the parameter determining the initial amount of smoothing in our perturbation process.

Conclusions:

If θ_0 is taken sufficiently large and if G and its derivatives up to the third order are sufficiently small, then the perturbation process will produce a perturbed imbedding of $\mathfrak{M}$ that is C^k and induces a metric tensor on $\mathfrak{M}$ which differs by the amount G from the metric induced by the original imbedding.

Part C: Preparatory Metric Approximation

In Theorem 1 of Part B we have the means for making small changes in the metric induced by an imbedding. Here in Part C we learn how to arrange that only a small change is needed. This solves the imbedding problem for compact Riemannian manifolds.

The Addition Property

Suppose a manifold $\mathfrak{M}$ has two imbeddings, one by functions z^α into E^m and the other by functions y^β into E^p. Let

(C1)
$$g_z = \sum_\alpha \frac{\partial z_\alpha}{\partial x_i} \frac{\partial z_\alpha}{\partial x_j} \quad \text{and}$$
$$g_y = \sum_\beta \frac{\partial y_\beta}{\partial x_i} \frac{\partial y_\beta}{\partial x_j}$$

be the two metric tensors defined by the two imbeddings, both referred to a system $x_1, x_2, \cdots x_n$ of local coordinates in $\mathfrak{M}$. The total set of functions $z_1, z_2, \cdots z_m, y_1, y_2, \cdots y_p$ defines an imbedding of $\mathfrak{M}$ into the product space $E^m \times E^p$. The metric tensor induced on $\mathfrak{M}$ by this product imbedding is $g_z + g_y$. This is the addition property of metric tensors.

This property enables us to separate into two parts the problem of constructing an imbedding of $\mathfrak{M}$ such that we can successfully apply Theorem 1 and obtain an isometric imbedding. Suppose g is the intrinsic metric of $\mathfrak{M}$ which we want to realize by an imbedding. We first find a z-imbedding which is "perturbable", that is, one such that the equation system (B8a, b) is everywhere non-singular. Then we find a y-imbedding such that $g_z + g_y$ is close to g. The perturbation process is now applied only to the z-imbedding, directed towards effecting the change $G = g - (g_z + g_y)$ in the metric induced by the z-imbedding.

Actually, the form of Theorem 1 forces us to proceed somewhat like this. This theorem tells us that for any perturbable imbedding a sufficiently small change G in the metric can be accomplished. To use it we must be able to make G small without changing the imbedding which is to be perturbed. And derivatives of G up to third order must be small. Thus it is rather important that we make $G(= g - g_z - g_y)$ small by adjusting the y-imbedding, leaving the z-imbedding fixed.

The metric g_y must be a positive metric, so it is clear that it would be impossible to use the approach we outlined above unless $g - g_z$ were a positive metric, which g_y would approximate to. If we have a z-imbedding that is perturbable (i.e. where (B8a, b) is everywhere a non-singular system) we can always simply make a change of scale, if necessary, to make g_z as small as desired so that $g - g_z$ will be positive. This does not affect the quite qualitative question of the singularity or non-singularity of (B8a, b).

A Simplified Approach

The method we actually use here for constructing the z and y imbeddings has a certain intricacy occasioned by our desire to bound the number of dimensions needed for the final imbedding and to get a relatively good bound. But if one throws out all concern for the number of dimensions to be used, the problem can be handled rather simply. Therefore we indicate here the simpler approach for the benefit of those who may not want to bother with the more complicated details of our method.

One could get by with two lemmas:

L_1. *Every compact differentiable manifold may be imbedded as an analytic sub-manifold of some euclidean space in such a way that* (B8a, b) *is everywhere non-singular.*

L_2. *Any compact Riemannian manifold with C^k positive metric, where $3 \leqq k \leqq \infty$, can be represented as an analytic sub-manifold of euclidean space so that the induced metric and its derivatives up to the third order approximate the given metric as closely as desired.*

These lemmas would not give us any indication of the number of dimensions necessary.

L_1 can be proved very easily by a direct construction. Take an analytic imbedding of $\mathfrak{M}$ in E^{2n}, where $n = \dim \mathfrak{M}$. Let $v_1, v_2, \cdots, v_{2n}$ be the coordinates of E^{2n}. Then the $2n^2 + 3n$ functions $v_1, v_2, \cdots, v_{2n}$; $v_1^2, v_2^2, \cdots, v_{2n}^2$; $v_1 v_2, v_1 v_3, \cdots, v_2 v_3, \cdots$ define an imbedding of $\mathfrak{M}$ in $(2n^2 + 3n)$-space if they are used as imbedding functions. This imbedding has the property we want, that (B8a, b) is everywhere nonsingular. To verify this, consider n of the v's as local coordinates at a point of $\mathfrak{M}$. [See (C7).]

L_2 could be proved by considering finite dimensional approximations to a C^k isometric imbedding of $\mathfrak{M}$ in Hilbert space. Or one could use a result of J. Schwartz that implies L_2. It is relatively easy to patch up a good approximation to a metric when there is no limit on the number of imbedding functions to be used. Each function might vanish except on a small neighborhood of $\mathfrak{M}$.

Outline of the Method

Our method for constructing y-imbeddings can use a minimum number of dimensions if the z-imbedding is chosen with some care. It turns out that if $g - g_z$ is close to a favorable metric γ the y-imbedding can be made in $n^2 + 3n$ dimensions.

Our first step is to find a favorable metric γ. Then we construct a z-imbedding such that $g - g_z$ approximates γ (derivatives are not involved here). Finally we construct the y-imbedding so that g_y approximates $g - g_z$ and its derivatives.

Determination of γ

The special mechanism we use in constructing the y-imbedding requires a set $\{\psi^r\}$ of functions on $\mathfrak{M}$ such that the set of symmetric tensors

$$(C2) \qquad M_{ij}^r = \frac{\partial \psi^r}{\partial x_i} \frac{\partial \psi^r}{\partial x_j}$$

has at every point of $\mathfrak{M}$ a subset of $\frac{1}{2}n^2 + \frac{1}{2}n$ tensors that are linearly independent. The x_i's are local coordinates in $\mathfrak{M}$. The metric γ will be the sum over r of the M_{ij}^r, or in other words, it will be the metric induced by the imbedding defined by using the ψ^r as imbedding functions.

Our construction of the ψ^r is based on a dimensionality argument. If we arrange to deal exclusively with algebraic functions and conditions we can think in terms of the precise dimensionality concepts of algebraic geometry (such as those based on transcendance degree). Therefore let $\mathfrak{A}$ be an algebraic representation[2] of $\mathfrak{M}$ in E^α. The mapping between $\mathfrak{M}$ and $\mathfrak{A}$ can be as differentiable as the differentiability structure on $\mathfrak{M}$, and analytic if $\mathfrak{M}$ has a *global* analytic structure (*not* mere overlapping local coordinates analytically related). Then a C^k metric on $\mathfrak{M}$ becomes a C^k metric on $\mathfrak{A}$.

Let us first see how we can find a set of functions which has the independence or "(C2) property" in the neighborhood of a point of $\mathfrak{A}$. Suppose $x_1, x_2, \cdots, x_n$ are local coordinates. Then the $\frac{1}{2}n^2 + \frac{1}{2}n$ functions

$$\text{(C3)} \qquad\qquad f^{ij} = x_i + x_j,$$

where $i \leqq j$, can easily be seen to suffice. At any point of $\mathfrak{A}$ some n of the coordinates $u_1, u_2, \cdots, u_a$ of E^α will be suitable as local coordinates. Therefore the $\frac{1}{2}a^2 + \frac{1}{2}a$ functions

$$\text{(C4)} \qquad\qquad f^{\beta\delta} = u_\beta + u_\delta,$$

where $\beta \leqq \delta$, will have the "(C2) property" everywhere. But this is more functions than we want to use.

A plausibility argument for the number of functions ψ^r that should be necessary so that there are always $\frac{1}{2}n^2 + \frac{1}{2}n$ linearly independent M^r_{ij} at a point of $\mathfrak{A}$ goes as follows: $\frac{1}{2}n^2 + \frac{1}{2}n$ functions are good locally, so they would fail on a sub-manifold of $n - 1$ dimensions. Adding a total of $n - 1$ functions should reduce it step by step to a zero-dimensional set of singular points. Then one more function should eliminate these. Thus we should need $\frac{1}{2}n^2 + 1\frac{1}{2}n$ functions in all.

$\frac{1}{2}n^2 + 1\frac{1}{2}n$ is the correct number, although that argument is not rigorous. Define $\frac{1}{2}n^2 + 1\frac{1}{2}n$ functions

---

2. Reference [16] proves that there is an algebraic representation of any closed differentiable manifold.

$$(C5) \qquad \psi^r = \sum_\beta C_\beta^r u_\beta \left[\begin{array}{l} r = 1, 2, \cdots, \frac{1}{2} n^2 + 1\frac{1}{2} n \\ \beta = 1, 2, \cdots, a \end{array} \right]$$

as linear combinations of the coordinates of E^a. We shall show that a generic choice of the coefficients C_β^r automatically gives a set of ψ^r with the desired property. Let $s = \frac{1}{2} n^2 + \frac{1}{2} n$. Then there are $s + n$ of the ψ^r and $(s + n)a$ of the coefficients C_β^r.

Our dimensionality argument is based on analyzing the family of ways in which a set of the ψ^r can fail to define independent M_{ij}^r. If the M_{ij}^r are not linearly independent at a point p of $\mathfrak{A}$ they lie in some linear sub-space H_p of the space L_p of all values of symmetric tensors (with two subscripts) at p. We can consider only sub-spaces H_p which have one less dimension than the whole linear space L_p. Since dim $L_p = s$, dim $H_p = s - 1$ and the dimension of the family of sub-spaces H_p of L_p is $s - 1$. The dimension of the family of all H_p for all points p of $\mathfrak{A}$ is $n + s - 1$.

For any r coefficients $(C_1^r, C_2^r, \cdots, C_a^r)$ can be chosen so that ψ^r is any one of the functions $f^{\beta\delta}$ of (C4). Therefore for any particular H_p, one can select $(C_1^r, C_2^r, \cdots, C_a^r)$ so that M_{ij}^r (which they determine, via ψ^r) does not lie in H_p. If this were not so the $f^{\beta\delta}$ would not have the "(C2) property," but we saw that they did have this property. Since not all selections of the constants determining ψ^r make M_{ij}^r lie in H_p, the dimension of the family of selections of $(C_1^r, C_2^r, \cdots, C_a^r)$ that do make $M_{ij}^r \varepsilon H_p$ is not more than $a - 1$.

The dimension of the family of selections of all the C_β^r which make all the M_{ij}^r lie in H_p is clearly not more than $(s + n)(a - 1)$ since there are $s + n$ sets of a coefficients determining the $s + n$ functions ψ^r which determine the M_{ij}^r. Now since the family of all H_p has dimension $n + s - 1$, the dimension of the family of selections of the C_β^r for which there is some H_p such that all the M_{ij}^r lie in H_p at p can be at most $[(s + n)(a - 1)] + [n + s - 1]$. But this number is $(s + n)a - 1$, one less than the dimension of all selections of the C_β^r. Consequently with a generic selection of the C_β^r there will be no point p and tensor sub-space H_p where all the M_{ij}^r lie in H_p. In other terms, the ψ^r will have the "(C2)", or independence, property we want.

One can be quite explicit about how the C_β^r can be chosen. The

n-dimensional variety associated with $\mathfrak{A}$ (which is the smallest variety containing $\mathfrak{A}$) can be defined by a set of polynomial equations in the coordinates of E^a. Adjoin all the coefficients involved in these equations to the field of rationals to produce an extension F. If the C_β^r are algebraically independent over F the ψ^r will have the desired property.[3] Obviously this is a sufficient but not necessary condition for the proper selection of the C_β^r.

We have said that the "favorable metric" γ would be the sum of the M_{ij}^r, so that

(C6)
$$\gamma_{ij} = \sum_\gamma M_{ij}^r = \sum_\gamma \frac{\partial \psi^r}{\partial x_i} \frac{\partial \psi^r}{\partial x_j}.$$

We also said that our procedure was to make $g - g_z \approx \gamma$. This means $g_z \approx g - \gamma$, so we must certainly have $g - \gamma$ a positive metric. To take care of this let us assume that a definite choice of the C_β^r is made in a such a way that they are small enough to make $g - \gamma$ be a positive metric.

The Z-Imbedding

There will be a C^1 imbedding of $\mathfrak{M}$ in E^{2n} which realizes the metric $g - \gamma$ exactly [9]. We can think of this as a mapping from $\mathfrak{A}$ into E^{2n} and approximate it in the C^1 sense by an algebraic imbedding $\mathfrak{B}$. Let g_b be the metric induced by this imbedding $\mathfrak{B}$. The approximation between g_b and $g - \gamma$ can be as close as desired, but it does not extend to derivatives.

Let $v_1, v_2, \cdots, v_{2n}$ be the coordinates of E^{2n}. At any point of $\mathfrak{B}$ some n of these will be suitable as local coordinates; let $x_1, x_2, \cdots, x_n$ be this subset.

Now consider the functions

$$z_i = x_i$$
$$z_{ij} = x_i x_j,$$

(C7)   where

$$i \leqq j.$$

---

3. By using rational coefficients in the approximating polynomials used in [16] one would obtain an algebraic imbedding $\mathfrak{A}$ such that the equations defining the corresponding variety would have rational coefficients. Then it would suffice to simply select the C_β^r as independent transcendentals over the field of rationals, without reference to $\mathfrak{A}$.

If these are the functions of z_α of (B8a, b) that system takes a very simple form:

$$\sum_\alpha \frac{\partial z_\alpha}{\partial x_i} \dot{z}_\alpha = \dot{z}_i = 0$$

$$-2\sum_\alpha \frac{\partial^2 z_\alpha}{\partial x_i \partial x_j} \dot{z}_\alpha = -2\dot{z}_{ij} = \dot{g}_{ij}.$$

The solution is apparent for any $\dot{g}_{ij}$. The system is non-singular in the most obvious way. Each linear equation has just one variable (remember the $\dot{z}_\alpha$ are the variables) with non-vanishing coefficient and this is a different variable in each equation.

The system (B8a, b) has the property that once one has a set of functions z_α that makes the system non-singular then the introduction of new functions z_α (which also introduces new variables $\dot{z}_\alpha$) can only improve the situation. The coefficients of each variable $\dot{z}_\alpha$ in the equations can be regarded as a vector V_α with $\frac{1}{2}n^2 + \frac{1}{2}n + n$ or $\frac{1}{2}n^2 + 1\frac{1}{2}n$ components. The system is non-singular if there are $\frac{1}{2}n^2 + 1\frac{1}{2}n$, which is $s + n$, linearly independent V_α's. Additional functions z_α simply introduce more V_α's.

Define $s + 2n$ (or $\frac{1}{2}n^2 + 2\frac{1}{2}n$) quadratic functions of the coordinates of E^{2n} by

(C8) $$z_\alpha = \sum_{1 \leq \beta \leq 2n} C_\alpha^\beta v_\beta + \sum_{1 \leq \beta \leq \delta \leq 2n} D_\alpha^{\beta\delta} v_\beta v_\delta,$$

where $1 \leq \alpha \leq s + 2n$. At any point of $\mathfrak{B}$ we can make the z_α contain the functions defined by (C7) which are suitable at that point by an appropriate choice of the C's and the D's. An argument exactly analogous to the one we used above in finding the ψ^r shows that a generic choice of the C's and D's gives functions z_α that make (B8a, b) non-singular. If the C's and D's are algebraically independent over the field of definition for $\mathfrak{B}$ they define satisfactory functions z_α.

Now how do we arrange that $g_z \approx g - \gamma$? We select all $D_\alpha^{\beta\delta} \approx 0$ and $C_\alpha^\beta \approx 0$ unless $\beta = \alpha$, when we select $C_\alpha^\alpha \approx 1$. This makes the z-imbedding, which is in E^{s+2n}, be approximately the same as $\mathfrak{B}$, because all the z^α are quite small, except for the first $2n$, and these are approximately the coordinates $v_1, v_2, \cdots, v_{2n}$ of E^{2n}. Thus $g_z \approx g_b$ and hence $g_z \approx g - \gamma$ since $g_b \approx g - \gamma$.

The Y-Imbedding

This is constructed, with the aid of the ψ^r of (C2), by means of a special device. The device will produce an imbedding with metric g_y approximating any metric of the form

(C9) $$g(a_1, a_2, \cdots, a_{s+n}) = \sum_r a_r M_{ij}^r,$$

where the M_{ij}^r are those of (C2) and the a_r are positive analytic functions on $\mathfrak{M}$. (Regard $\mathfrak{M}$ as having an analytic structure corresponding to those of its imbeddings $\mathfrak{A}$ and $\mathfrak{B}$.) This approximation will apply to derivatives also.

Note that $g(1, 1, \cdots 1) = \gamma$. Since the M_{ij}^r are linearly independent at each point and $g - g_z \approx \gamma$, we can certainly represent $g - g_z$ in the form

(C10) $$g - g_z = \sum_r \alpha_r M_{ij}^r$$

at any point of $\mathfrak{M}$, and we can have $|\alpha_r - 1| \leq \varepsilon$ where ε is a small uniform bound that depends on the closeness of the approximation of $g - g_z$ to γ.

But can we do this in a uniform manner so that the α_r become continuous functions on $\mathfrak{M}$? The solution for the α_r of (C10) will be non-unique at every point because there are $s + n$ of the α_r and only s components of $g - g_z$, thus only s equations. This redundancy can be removed by the same device which removed the redundancy of (B8a, b) in Part B. Here we desire positive α_r, so we specify:

(C11) $$\sum_r (\alpha_r - 1)^2 = \text{ minimum, subject to (C10).}$$

This determines the α_r uniquely and makes them have the same differentiability as $g - g_z$. Since (C10) has a solution with $|\alpha_r - 1| \leq \varepsilon$, the solution of the modified system (C10, 11) will have $|\alpha_r - 1| \leq \epsilon(s+n)^{\frac{1}{2}}$. Assume ε is small enough so that this makes the α_r necessarily positive.

We define $g(a_1, a_2, \cdots a_{n+s})$ by approximating the α_r by positive analytic functions a_r. The a_r must also approximate the α_r for derivatives up to the third order. Hence $g - g_z$ must be C^3 so that the α_r are C^3. Since $a_r \approx_3 \alpha_r$ we shall have $g(a_1, a_2, \cdots, a_{n+s}) \approx_3 g - g_z$.

The Device

This device is reminiscent of one used for constructing C^1 imbeddings.[4] Suppose λ is a large constant. We define $2(s+n)$, which is $n^2 + 3n$, y-imbedding functions as

(C12)
$$y_r = \frac{(a_r)^{\frac{1}{2}}}{\lambda} \sin(\lambda \psi^r)$$

$$\bar{y}_r = \frac{(a_r)^{\frac{1}{2}}}{\lambda} \cos(\lambda \psi^r).$$

When these are used as imbedding functions they induce the metric

$$g_y = \sum_r \frac{\partial y_r}{\partial x_i} \frac{\partial y_r}{\partial x_j} + \sum_r \frac{\partial \bar{y}_r}{\partial x_i} \frac{\partial \bar{y}_r}{\partial x_j}.$$

When this is expanded by substituting the formulas for y_r and $\bar{y}_r$ many terms cancel. All terms which contain λ^{-1} occur as pairs which differ only by containing either $[\sin(\lambda \psi^r)] \cdot [\cos(\lambda \psi^r)]$ or $[\cos(\lambda \psi^r)] \cdot [-\sin(\lambda \psi^r)]$ and cancel together. The remaining terms can be combined with the identity $\sin^2 + \cos^2 = 1$. This finally gives

$$g_y = \sum_r a_r \frac{\partial \psi^r}{\partial x_i} \frac{\partial \psi^r}{\partial x_j} + \lambda^{-2} \sum_r \frac{\partial(a_r)}{\partial x_i} \frac{\partial(a_r)}{\partial x_j}, \quad \text{or}$$

$$g_y = \sum_r a_r M_{ij}^r + \lambda^{-2} \bar{g}$$

$$= g(a_1, \cdots, a_{s+n}) + \lambda^{-2} \bar{g}.$$

Now $\bar{g}$ is an analytic tensor independent of λ. By choosing λ very large the error $\lambda^{-2} \bar{g}$ and any number of its derivatives can be made as small as desired. So we have

(C14)
$$g_y \approx_3 g(a_1, a_2, \cdots, a_{s+n}) \approx_3 g - g_z,$$

where $\approx_3$ indicates approximation up to third derivatives. Thus we have $g_y + g_z \approx_3 g$, which is what we need to apply Theorem 1. This requires, of course, that g be C^3.

Summary and Applications

We used $s + 2n$, which is $\frac{1}{2}n^2 + 2\frac{1}{2}n$, z-imbedding functions and

---

4. See equation (13) page 387 of [9].

$2(s + n)$, or $n^2 + 3n$, y-imbedding functions. This is $3s + 4n$ or $1\frac{1}{2}n^2 + 5\frac{1}{2}n$ functions altogether. The z-imbedding was analytic and made (B8a, b) everywhere non-singular, so that Theorem 1 could be applied to it. The y-imbedding was also analytic and was adjustable so that $g_z + g_y$ could approximate g as closely as desired (this approximation including derivatives up to the third order). For this, g had to be C^3.

The z- and y-imbeddings can be arranged to take up arbitrarily little space. The z-imbedding approximates a C^1 imbedding of $\mathfrak{M}$ which realizes the metric $g - \gamma$. Since a C^1 isometric imbedding can be made arbitrarily small (and highly twisted) so can the z-imbedding. If the parameter λ is very large the y-imbedding is very small.

The fact that the z-imbedding approximates a C^1 imbedding serves to prevent self-intersections in the final imbedding of the manifold in $E^{1\frac{1}{2}n^2+5\frac{1}{2}n}$. Since the amount of perturbation needed can be made arbitrarily small by adjusting the y-imbedding, the application of the perturbation process to the z-imbedding need not produce self-intersections.

Now we state the result obtained by combining the work of Part C with Theorem 1 of Part B. This is our "main theorem."

Imbedding of Compact Manifolds

THEOREM 2. *A compact n-manifold with a C^k positive metric has a C^k isometric imbedding in any small volume of euclidean $(n/2)(3n + 11)$-space, provided $3 \leqq k \leqq \infty$.*

Part D: Non-Compact Manifolds

Our treatment here is not a direct attack. It exploits a device by which the imbedding problem for non-compact manifolds is reduced to the problem for compact manifolds. This approach gives a poor upper bound on the number of dimensions needed for the imbedding space; but that is the price of taking a shortcut to the non-compact case.

A Special Mapping

Our basic tool is a C^∞ mapping of E^n onto the n-sphere S^n. Most of E^n is mapped into the "north pole" of S^n. The interior of the unit disk of E^n covers the remaining portion of S^n in a one-to-one manner. Any mapping of E^n on S^n with these properties will serve our purpose, but to illustrate we construct one.

Take the case of the (x, y) plane, or E^2. This can be mapped on the 2-sphere $\xi^2 + \eta^2 + \zeta^2 = \frac{1}{4}$ as follows:

For $x^2 + y^2 < 1$, let

$$Q = \exp\left(x^2 + y^2 - 1\right)^{-1}$$

$$\xi = \frac{xQ}{Q^2 + x^2 + y^2}$$

$$\eta = \frac{yQ}{Q^2 + x^2 + y^2}$$

$$\zeta = \frac{1}{2} - \frac{Q^2}{Q^2 + x^2 + y^2};$$

for $x^2 + y^2 \geqq 1, \xi = \eta = 0, \zeta = \frac{1}{2}$.

It is easy to see that ξ, η, and ζ are C^∞ functions because Q is a C^∞ function if it is assigned the value zero for $x^2 + y^2 \geqq 1$. A direct check verifies that

$$\xi^2 + \eta^2 + \zeta^2 = \frac{1}{4}.$$

The equations define a non-singular one-to-one mapping of the open disk $x^2 + y^2 < 1$ onto the sphere minus the "north pole" ($\xi = \eta = 0, \zeta = \frac{1}{2}$). This mapping is obtainable by taking a mapping of the open disk onto the whole plane,

$$\bar{x} = x/Q$$
$$\bar{y} = y/Q$$

and following this by the classical conformal mapping of the plane onto the sphere (minus the "north pole"). The effect is to give a mapping which has a C^∞ extension to the rest of the plane where all points not interior to the disk map into the "north pole."

Patch Mappings

Consider a C^∞ Riemannian n-manifold $\mathfrak{M}$ (the metric need not be C^∞, but $\mathfrak{M}$ has a C^∞ structure). A local coordinate system or neighborhood

N in $\mathfrak{M}$ can be regarded as the image of the unit disk D of E^n under a C^∞ mapping of D into $\mathfrak{M}$. This mapping should be one-to-one, non-singular, and extensible to an open set of E^n containing D, as a C^∞ non-singular mapping. Then an open set containing N is mapped on the open set containing D by the C^∞ inverse mapping.

For brevity call S the n-sphere. Our special mapping will map any open set containing D onto S in a C^∞ manner. This mapping with the inverse mapping $N \to D$, gives a mapping

$$N \overset{\varphi}{\to} S$$

which is C^∞ and has a C^∞ extension to an open set containing N. φ maps all points on the boundary of N or outside N into the "north pole" of S. Clearly we can extend the definition of φ to all points of $\mathfrak{M}$ by mapping all other points into this "north pole". φ will remain C^∞. This mapping φ has non-vanishing Jacobian in the interior of N, so φ^{-1} is C^∞ there. φ is called a patch mapping.

Appropriate Coverings for $\mathfrak{M}$

$\mathfrak{M}$ can be covered by a family of disk neighborhoods N_i in such a way that we can divide the N_i among $n + 1$ classes where: No two N_i of the same class overlap. Each N_i overlaps only a finite number of other N_i.

How is such a covering constructed? First obtain a regular star-finite cellular sub-division of $\mathfrak{M}$. Then form a disk neighborhood corresponding to each vertex, edge, . . . , face, or cell of the cellular sub-division. Each disk neighborhood that corresponds (for example) to an edge covers the middle section of the edge but not the end points. These are covered by the neighborhoods which correspond to them in their role of vertices. In this way no two edge neighborhoods are allowed to meet. The same principle applies up the series of dimensions. The $n + 1$ dimensions from 0 through n give rise to the $n + 1$ classes of neighborhoods.

Within each N_i we can select a slightly smaller disk neighborhood $\bar{N}_i$ such that the $\bar{N}_i$ also cover $\mathfrak{M}$. These $\bar{N}_i$ should correspond to sub-disks of D (through the mappings between D and the N_i). Then we can select a C^∞ function u_i for each $\bar{N}_i$ which is positive interior to $\bar{N}_i$

and zero on the boundary and outside $\bar{N}_i$. Each u_i can be regarded as defined and C^∞ over all of $\mathfrak{M}$.

Now if we define

$$v_i = u_i \Big/ \sum_i u_i$$

the v_i form a partition of unity by C^∞ functions, each of which is positive interior to the corresponding sub-neighborhood $\bar{N}_i$ and zero everywhere else.

Assignment of Metrics

Each N_i has an associated patch mapping

$$N_i \xrightarrow{\varphi_i} S_i.$$

We write S_i to distinguish different n-spheres for different N_i. The mapping φ_i has a C^∞ non-singular inverse φ_i^{-1} on $\bar{N}_i$.

Consider a metric γ_{i0} on S_i. This gives a corresponding metric g_{i0} on N_i. Actually g_{i0} can be regarded as defined for all $\mathfrak{M}$ because it will be zero at the boundary of N_i and can be extended by defining it as zero on the rest of $\mathfrak{M}$. If it is C^k on S_i it will be C^k on $\mathfrak{M}$ also. If we select a metric γ_{i0} on each S_i which is positive, C^∞, and sufficiently small the corresponding metrics g_{i0} will add on $\mathfrak{M}$ to a metric

$$g_0 = \sum_i g_{i0}$$

which is C^∞ and everywhere shorter than g, the metric we are given on $\mathfrak{M}$, and which we want to realize by an imbedding of $\mathfrak{M}$. For example, each γ_{i0} could be the metric induced by an imbedding of S_i in E^{n+1} as a small geometrical sphere.

Now, since $g - g_0$ is a positive metric,

$$g_i = g_{i0} + v_i(g - g_0)$$

will be positive and will be as differentiable as g (say C^k). g_i differs from g_{i0} only within $\bar{N}_i$ where the mapping φ_i has a non-singular inverse φ_i^{-1}. Therefore φ_i^{-1} carries $g_i - g_{i0}$ over to S_i as a C^k non-negative metric $\gamma_i - \gamma_{i0}$. That is, there is a positive C^k metric γ_i on S_i which corresponds, via φ_i, to the metric g_i on N_i.

Consider the sum

$$\sum_i g_i = \sum_i g_{i0} + \left(\sum_i v_i\right)(g - g_0)$$
$$= g_0 + (g - g_0)$$
$$= g.$$

Realization of Metrics

We have defined a C^k metric γ_i on each S_i. Assuming $3 \leqq k \leqq \infty$, γ_i can be realized by a C^k imbedding of S_i in $E^{(n/2)(3n+11)}$ by our Theorem 2. We can always let the "north pole" of S_i be at the origin.

Now consider all the N_i of one of the $n + 1$ classes, let us say class C. The S_i corresponding to each has an imbedding in $E^{(n/2)(3n+11)}$ that realizes γ_i and maps the "north pole" into the origin. The corresponding patch mappings φ_i, together with these imbeddings, define a mapping ψ_c of $\mathfrak{M}$ into $E^{(n/2)(3n+11)}$. This ψ_c is C^∞ and maps all points of $\mathfrak{M}$, except those interior to any neighborhood N_i of class C, into the origin. Each of these other points can be in only one N_i, so the mapping is unambiguously defined.

This mapping ψ_c induces a metric on $\mathfrak{M}$ which is the sum,

$$g_c = \sum_{N_i \in C} g_i,$$

of the metrics g_i associated with neighborhoods N_i of class C. The product mapping

$$\psi = \psi_1 \times \psi_2 \times \cdots \times \psi_{n+1}$$

maps $\mathfrak{M}$ into $(n + 1)(n/2)(3n + 11)$ dimensional space and is also C^k. It induces the metric

$$g = \sum_c g_c,$$

as we desired. If the image of $\mathfrak{M}$ under ψ has no self-intersections it is an isometric imbedding. In any case it is a C^k isometric immersion.

Avoidance of Self-Intersections

Self-intersections can be avoided by using the fact that the isometric imbedding of S_i can be made as small as desired. Let α_i be the minimum distance of a point of $\bar{N}_i$ from the origin after it is mapped into $E^{1\frac{1}{2}n^2+5\frac{1}{2}n}$ by φ_i and the imbedding of S_i. Let β_i be the maximum distance from the origin of points of $\bar{N}_i$ after being mapped into $E^{1\frac{1}{2}n^2+5\frac{1}{2}n}$. Think of i as a serial index running 1, 2, 3. . . . . Now all we need do to avoid self-intersections is to arrange the imbeddings of the S_i in order of increasing i so that for all i

$$\beta_i < \min_{j<i} \alpha_j.$$

Why is this sufficient? First, any two points of $\mathfrak{M}$ that are interior to a common N_i are distinguished through the imbedding of S_i. So we need only consider pairs of points which lie in completely different sets of neighborhoods. In this case one of the points will be in a subneighborhood $\bar{N}_i$ with lower index than the other. Then this point will be further from the origin with respect to the set of $1\frac{1}{2}n^2 + 5\frac{1}{2}n$ coordinates associated with N_i than the other point can be.

The Result

Our theorem is:

THEOREM 3. *Any Riemannian n-manifold with C^k positive metric, where $3 \leqq k \leqq \infty$, has a C^k isometric imbedding in $(1\frac{1}{2}n^3+7n^2+5\frac{1}{2}n)$-space; in fact, in any small portion of this space.*

Bibliography

1. C. F. Gauss, Disquisitiones generales circa superfices curvas, Werke IV, 1827.
2. B. Riemann, *Über die hypotesen welche der geometrie zugrunde liegen*, Gött. Abh., 13 (1868), pp. 1–20.
3. L. Schlaefli, *Nota alla memoria del. Sig. Beltrami, Sugli spazii di curvatura constante*, Ann. di mat., 2^e serie, 5 (1871–1873), pp. 170–93.
4. D. Hilbert, *Ueber flächen von constanter Gausscher krümmung*, Trans. Amer. Math. Soc., 2 (1901), pp. 87–99.
5. C. Tompkins, *Isometric embedding of flat manifolds in Euclidean space*, Duke Math. J., 5 (1939), pp. 58–61.
6. S. S. Chern and N. H. Kuiper, *Some theorems on the isometric imbedding of compact Riemann manifolds in Euclidean space*, Ann. of Math., 56 (1952), pp. 422–30.
7. M. Janet, *Sur la possibilité de plonger un espace riemannien donné dans un espace euclidien*, Annales de la société polonaise de mathématique, 5 (1926), pp. 38–43.

8. E. Cartan, *Sur la possibilité de plonger un espace riemannien donné dans un espace euclidien,* Annals de la société polonaise de mathématique, 6 (1927), pp. 1–7.

9. J. Nash, C^1 *isometric imbeddings,* Ann. of Math., 60 (1954), pp. 383–96.

10. N. H. Kuiper, *On* C^1 *isometric imbeddings,* Proc. Kon. Ac. Wet. Amsterdam A 58 (Indigationes Mathematicae), no. 4 (1955), pp. 545–56.

11. H. Weyl, *Über die bestimmung einer geschlossenen konvexen fläche durch ihr linienelement,* Vierteljahrsschrift der naturforschender Gesellschaft, Zurich, 61 (1916), pp. 40–72.

12. H. Lewy, *On the existence of a closed convex surface realizing a given Riemannian metric,* Proc. Nat. Acad. Sci. U.S.A., 24, No. 2 (1938), pp. 104–6.

13. A. D. Alexandrov, Intrinsic geometry of convex surfaces, OGIZ, Moscow-Leningrad, 1948.

14. A. V. Pogorelov, Deformation of convex surfaces, Gosudarstv. Izdat. Tehn.–Teor. Lit., Moscow-Leningrad, 1951. (Also see later papers of Alexandrov and Pogorelov.)

15. Louis Nirenberg, *The Weyl and Minkowski problems in differential geometry in the large,* Comm. Pure Appl. Math., 6 (1953), pp. 337–94.

16. J. Nash, *Real algebraic manifolds,* Ann. of Math., 56 (1952), pp. 405–21.

17. Danilo Blanuša, *Über die Einbettung hyperbolischer Räume in euklidische Räume,* Monatshefte für Mathematik, 59 Band, 3 Heft (1955), pp. 217–29.

Author's Note to "The Imbedding Problem for Riemannian Manifolds"

In June 1998 I was notified by an e-mail from Professor R. M. Solovay of a fault in the arguments of the last part (part D) of my paper "The Imbedding Problem for Riemannian Manifolds." Later in 1998 I put a reference to this error on my Web page as "erratum.txt".

It seems, surprisingly, that before then no reader had actually detected the error!

With regard to the question of repair or repairs of the error, I feel that the whole issue of what to do for non-compact manifolds has been changed by the contributions of Mikhail Gromov. It was in the nature of Gromov's methods that results were achieved for the embedding of non-compact manifolds with the same requirements on dimensions as those for compact manifolds.

So it becomes inappropriate to "patch up" my argument in the original part D to be sure that self-intersections are avoided and the result is definitely an embedding rather than merely an immersion. I claim that that could be done without needing more dimensions than I had specified for the flawed argument. But it is better to proceed with Gromov and thus need many fewer dimensions.

A good reference to Gromov's work is his book *Partial Differential Relations* (Berlin: Springer-Verlag, 1986).

CHAPTER 12

JOHN   F.   NASH,   JR.

Continuity of Solutions of Parabolic and Elliptic Equations

Introduction

Successful treatment of non-linear partial differential equations generally depends on "a priori" estimates controlling the behavior of solutions. These estimates are themselves theorems about linear equations with variable coefficients, and they can give a certain compactness to the class of possible solutions. Some such compactness is necessary for iterative or fixed-point techniques, such as the Schauder-Leray methods. Alternatively, the a priori estimates may establish continuity or smoothness of generalized solutions. The strongest estimates give quantitative information on the continuity of solutions without making quantitative assumptions about the continuity of the coefficients.

The theory of non-linear elliptic equations in two independent variables is fairly well developed. (See [1] for a survey and bibliography.) An essential part is the a priori Hölder continuity estimate for solutions of uniformly elliptic equations, first provided by Morrey in 1938. All methods used to obtain this estimate have been quite special

Received May 26, 1958.

to two dimensions, utilizing, for example, complex analysis and quasi-conformal mappings (see [2]). The restriction to two variables has been due to this use of such special methods; except for the crucial a priori estimate, the theory is extensible (and in large part has been extended) to n dimensions and to parabolic equations. Our results fill this gap, and it should now be possible to build a general theory of non-linear parabolic and elliptic equations, free of dimension restrictions. Strictly speaking, our work needs some generalization to cover equations with lower order terms, systems, etc. This generalization can probably be accomplished fairly quickly.

In this paper, we consider linear parabolic equations of the form

$$\sum_{i,j} \partial \left[C_{ij}(x_1, x_2, \cdots, x_n, t)\partial T/\partial x_j \right] /\partial x_i = \partial T/\partial t, \quad \text{or}$$

(1)

$$\nabla \cdot (C(x, t) \cdot \nabla T) = T_t,$$

where the C_{ij} form a symmetric real matrix $C(x, t)$ for each point x and time t. We assume there are universal bounds $c_2 \geqq c_1 > 0$ on the eigenvalues of C so that any eigenvalue θ_v satisfies $c_1 \leqq \theta_v \leqq c_2$. This is the standard "uniform ellipticity" assumption. The continuity estimate for a solution $T(x, t)$ of (1) satisfying $|T| \leqq B$ and defined for $t \geqq t_0$ is

$$|T(x_1, t_1) - T(x_2, t_2)|$$

(2)
$$\leqq BA \left\{ \left[|x_1 - x_2|/(t_1 - t_0)^{\frac{1}{2}} \right]^\alpha \right.$$

$$\left. + \left[(t_2 - t_1)/(t_1 - t_0) \right]^{\frac{1}{2}\alpha/(1+\alpha)} \right\},$$

where $t_2 \geqq t_1 > t_0$. Here A and α are a priori constants which depend only on c_1 and c_2 and the space dimensions n. As a corollary of our results on parabolic equations, we obtain a continuity estimate for solutions of elliptic equations. If $T(x)$ satisfies $\nabla \cdot (C(x) \cdot \nabla T) = 0$ in a region R and the same bounds c_1 and c_2 limit the eigenvalues of $C(x)$, then

(3) $$|T(x_1) - T(x_2)| \leqq BA' \left(|x_1 - x_2|/d(x_1, x_2) \right)^{\alpha/(1+\alpha)},$$

where α is the α of (2) and A' is an a priori constant $A'(n, c_1, c_2)$, and where $|T| \leqq B$ in R and $d(x_1, x_2)$ is the lesser of the distances of the points x_1 and x_2 from the boundary of R.

Our paper is arranged in six parts, each concluding with the attainment of a result significant in itself. Detailed proofs are given and all the results presented in [14] are covered. An appendix states further results, including continuity at the boundary in the Dirichlet problem, a Harnack inequality, and other results, stated without detailed proofs.

General Remarks

The open problems in the area of non-linear partial differential equations are very relevant to applied mathematics and science as a whole, perhaps more so than the open problems in any other area of mathematics, and this field seems poised for rapid development. It seems clear, however, that fresh methods must be employed. We hope this paper contributes significantly in this way and also that the new methods used in our previous paper, [10], will be of value.

Little is known about the existence, uniqueness, and smoothness of solutions of the general equations of flow for a viscous, compressible, and heat conducting fluid. These are a non-linear parabolic system of equations. Also the relationship between this continuum description of a fluid (gas) and the more physically valid statistical mechanical description is not well understood. (See [11], [12], and [13]). An interest in these questions led us to undertake this work. It became clear that nothing could be done about the continuum description of general fluid flow without the ability to handle non-linear parabolic equations and that this in turn required an a priori estimate of continuity, such as (2).

Probably one should first try to prove a conditional existence and uniqueness theorem for the flow equations. This should give existence, smoothness, and unique continuation (in time) of flows, conditional on the non-appearance of certain *gross* types of singularity, such as infinities of temperature or density. (A gross singularity could arise, for example, from a converging spherical shock wave.) A result of this kind would clarify the turbulence problem.

The methods used here were inspired by physical intuition, but the ritual of mathematical exposition tends to hide this natural basis. For parabolic equations, diffusion, Brownian movement, and flow of heat or electrical charges all provide helpful interpretations. Moreover, to us,

Continuity of Solutions of Parabolic and Elliptic Equations

parabolic equations seem more natural than elliptic ones. It is certainly true in principle that the theory of parabolic equations includes elliptic equations as a specialization, and in applications an elliptic equation typically arises as the description of the steady state of a system which in general is described by a parabolic equation.

In our work, no difference at all appears between dimensions two and three. Only in one dimension would the situation simplify. The key result seems to be the moment bound (13); it opens the door to the other results. We had to work hard to get (13), then the rest followed quickly.

We are indebted to several persons and institutions in connection with this work, including Bers, Beurling, Browder, Carleson, Lax, Levinson, Morrey, Newman, Nirenberg, Stein and Wiener, the Alfred P. Sloan Foundation, the Institute for Advanced Study, M.I.T., N.Y.U., and the Office of Naval Research.

Part I: The Moment Bound

More than enough is known about linear parabolic equations with variable coefficients to insure the existence of well behaved solutions for equations of the form (1) if we make strong (qualitative) restrictions on the C_{ij} and restrict the class of solutions to be considered. (See [3] through [7].) Therefore we assume: (a) the $C_{ij}(x, t)$ are uniformly C^∞, (b) $C_{ij}(x, t) = \sqrt{c_1 c_2}\delta_{ij}$ (Kronecker delta) for $|x| \geqq r_o$, some large constant. We consider only solutions $T(x, t)$ bounded in x for each t for which the solution is defined, i.e. max $|T(x, t)|$ is finite.

Under these restrictions, any bounded measurable function $T(x, t_0)$ of x given at an initial time t_0 determines a unique continuation $T(x, t)$ defined for all $t \geqq t_0$ and C^∞ for $t > t_0$. Moreover, $T(x, t) \to T(x, t_0)$ almost everywhere as $t \to t_{0_z}$ and $\max_x |T(x, t)|$ is nonincreasing in t. It is also known that fundamental solutions, which we discuss below, exist and have the general properties we state (see [4], [7]).

After the a priori results are established, a passage to the limit can remove the restrictions on the C_{ij}. This is a standard device in the use of a priori estimates. The Hölder continuity (2) makes the family of solutions equicontinuous and forces a continuous limit (generalized) solution to exist. Furthermore, the maximum principle remains valid and with it the unique continuability of solutions bounded in space. The final result requires only measurability for the C_{ij} plus the uniform ellipticity condition; and the a priori estimates then hold for the generalized solutions.

The use of fundamental solutions is very helpful with equations of the form (1). Our work is built around step by step control of the properties of fundamental solutions and most of the results concern them directly. A fundamental solution $T(x, t)$ has a "source point" x_0 and "starting time" t_0 and is defined and positive for $t > t_0$. Also, $\int T(x, t)\, dx = 1$ for every $t > t_0$, where dx is the volume element in n-space. As $t \to t_0$, the fundamental solution concentrates around its source point; $\lim T(x, t)$ is zero unless $x = x_0$, in which case it is $+\infty$. Physically, a fundamental solution represents the concentration of a diffusant spreading from an initial concentration of unit mass per volume at x_0 at time t_0.

All fundamental solutions are conveniently unified in a "characterizing function" $S(x, t, \bar{x}, \bar{t})$. For fixed $\bar{x}$ and $\bar{t}$ and as a function of x and t, S is a fundamental solution of (1) with source point $\bar{x}$ and starting time $\bar{t}$. Dually, for fixed x and t, S is a fundamental solution of the adjoint equation: $\nabla_{\bar{x}} \cdot [C(\bar{x}, t) \cdot \nabla_{\bar{x}} S] = -\partial S/\partial \bar{t}$, where time runs backwards. This duality enables us to use estimates for fundamental solutions in two ways on S.

The dependence of a bounded solution $T(x, t)$ on bounded initial data $T(x, t_0)$ is expressible through S:

$$(4) \qquad T(x; t) \int S(x, t, \bar{x}, t_0)\, T(\bar{x}, t_0)\, d\bar{x};$$

in particular,

$$(5) \qquad S(x_2, t_2, x_0, t_0) \int S(x_2, t_2, x_1, t_1) S(x_1, t_1, x_0, t_0)\, dx_1.$$

These are standard relations. (5) reveals a reproductive property of fundamental solutions.

Now consider a special fundamental solution $T = T(x, t) = S(x, t, 0, 0)$ with source at the origin and starting time zero. Let

$$E = \int T^2\, dx,$$

then

$$E_t = 2 \int T T_t \, dx = 2 \int T \nabla \cdot (C \cdot \nabla T) \, dx = -2 \int \nabla T \cdot C \cdot \nabla T \, dx,$$

by integration by parts. For any vector V, we have $c_1 |V|^2 \leq V \cdot C \cdot V \leq c_2 |V|^2$; therefore

(6)
$$-E_t \geq 2c_1 \int |\nabla T|^2 \, dx.$$

With (6) and a lower bound for $\int |\nabla T|^2 \, dx$ in terms of E, we shall be able to bound E above, obtaining our first a priori estimate. To bound $\int |\nabla T|^2 \, dx$ we employ a general inequality valid for any function $u(x)$ in n-space. For our purposes, we assume u is smooth and well behaved at infinity. E. M. Stein gave us the quick proof which follows below.

The Fourier transform of $u(x)$ is

$$v(y) = (2\pi)^{-n/2} \int e^{ix \cdot y} u(x) \, dx.$$

This has the familiar property

$$\int |v|^2 \, dy = \int |u|^2 \, dx.$$

The transform of $\partial u / \partial x_k$ is $iy_k v$; hence

$$\int |\partial u / \partial x_k|^2 \, dx = \int y_k^2 |v|^2 \, dy,$$

and

$$\int |\nabla u|^2 \, dx = \sum_k \int (\partial u / \partial x_k)^2 \, dx = \int |y|^2 |v|^2 \, dy.$$

Finally,

$$|v| \leq (2\pi)^{-n/2} \int |e^{ix \cdot y}| \cdot |u| \, dx = (2\pi)^{-n/2} \int |u| \, dx;$$

therefore, for any $\rho > 0$, we have

(a)
$$\int_{|y| \leq \rho} |v|^2 \, dy \leq \left(\pi^{n/2} \rho^n / (n/2)! \right) \left\{ (2\pi)^{-n/2} \int |u| \, dx \right\}^2,$$

using the formula for the volume of an n-sphere. On the other hand,

(b)
$$\int_{|y| > \rho} |v|^2 \, dy \leq \int_{|y| > \rho} |y/\rho|^2 |v|^2 \, dy = \rho^{-2} \int |\nabla u|^2 \, dx.$$

If we choose the value of ρ minimizing the sum of the two bounds (a) and (b), we obtain a bound on $\int |v|^2 \, dy = \int |u|^2 \, dx$ in terms of $\int |u| \, dx$ and $\int |\nabla u|^2 \, dx$. Solved for $\int |\nabla u|^2 \, dx$, this is

$$\int |\nabla u|^2 \, dx \geqq (4\pi n/(n+2)) \left[(n/2)!/(1+n/2)\right]^{2/n}$$

$$\left[\int |u| \, dx\right]^{-4/n} \left[\int |u|^2 \, dx\right]^{1+2/n}$$

Applying the above inequality with $u = T$, remembering that $\int T \, dx = 1$, we obtain from (6)

$$- E_t \geqq k E^{1+2/n}.$$

This is the first use of a convention we now establish that k is a generic symbol for a priori constants which depend only on n, c_1, and c_2. Any two instances of k should be presumed to be different constants. Thus, from the above inequality, $(E^{-2/n})_t \geqq k$; hence $E^{-2/n} \geqq kt$ and

(7) $$E \leqq kt^{-n/2}.$$

We used above the qualitative fact $\lim_{t \to 0} E = \infty$.

From this first bound (7) and the identity (5), we obtain

$$T(x, t) = \int S(x, t, \bar{x}, t/2) S(\bar{x}, t/2, 0, 0) \, d\bar{x},$$

whence

$$(T(x, t))^2 \leqq \int \left[S(x, t, \bar{x}, t/2)\right]^2 \, d\bar{x} \cdot \int \left[S(\bar{x}, t/2, 0, 0)\right]^2 \, d\bar{x}$$

$$\leqq \left[k(t/2)^{-n/2}\right]^2.$$

Therefore

(8) $$T \leqq kt^{-n/2},$$

which is a pointwise bound, stronger than (7).

The key estimate controls the "moment" of a fundamental solution

$$M = \int rT \, dx = \int |x| T \, dx.$$

To prove $M \leqq kt^{\frac{1}{2}}$ is our first major goal. This is dimensionally the only possible form for a bound on M. The moment bound is essential to all subsequent parts of this paper.

We also define an "entropy."

$$(9) \qquad Q = -\int T \log T \, dx.$$

From (8),

$$Q \geq \int \min_x \left[-\log T \right] (T \, dx) \geq -\log \left(kt^{-n/2} \right) \int T \, dx,$$

hence

$$(10) \qquad Q \geq \pm k + \tfrac{1}{2} n \log t$$

because $\int T \, dx = 1$. The sharp result $Q \geq \tfrac{1}{2} n \log(4\pi \, ec_1 t)$ is obtainable from a more sophisticated argument.

Our derivation of a bound on M requires a lower bound on M in terms of Q as a lemma. This inequality, which is $M \geq ke^{Q/n}$, depends only on the facts $T \geq 0$, $\int T \, dx = 1$. First observe that for any fixed λ,

$$\min_T (T \log T + \lambda T) = -e^{-\lambda - 1}.$$

Let $\lambda = ar + b$, where $r = |x|$ and a and b are any constants, and integrate over space, obtaining

$$\int \left[T \log T + (ar + b) T \right] dx \geq e^{-b-1} \int e^{-ar} \, dx,$$

or

$$-Q + aM + b \geq -e^{-b-1} a^{-n} D_n,$$

where D_n is the well known constant $2^n \pi^{\frac{1}{2}(n-1)} \left[\tfrac{1}{2}(n-1) \right]!$ related to the gamma-function and the surface of the $(n-1)$-sphere. Now set $a = n/M$ and $e^{-b} = (e/D_n) \cdot a^n$. Then $-Q + n + b \geq -1$ or $n + 1 \geq Q + \log(n/D_n) + \log(n/M)$; thus $n \log M + n \geq Q + n \log n - \log D_n$, finally,

$$(11) \qquad M \geq \left(n/eD_n^{1/n} \right) e^{Q/n} = ke^{Q/n}.$$

This ingenious proof, due to L. Carleson, gives an optimal constant.

The next inequality is a "dynamic" one, connecting the rates of change with time of M and Q. Differentiating (9),

$$Q_t = -\int (1 + \log T)\, T_t \, dx = -\int (1 + \log T)\nabla \cdot (C \cdot \nabla T)\, dx$$

$$= \int \nabla(\log T) \cdot C \cdot \nabla T \, dx$$

after integration by parts. This can be rewritten

$$Q_t = \int \nabla(\log T) \cdot C \cdot \nabla(\log T)(T \, dx).$$

Since in general $V \cdot c_2 C \cdot V \geqq V \cdot C^2 \cdot V = |C \cdot V|^2$, where V is a vector, we have

$$c_2 Q_t \geqq \int |C \cdot \nabla(\log T)|^2 (T \, dx) \geqq \left[\int |C \cdot \nabla \log T|(T \, dx)\right]^2$$

$$\geqq \left[\int |C \cdot \nabla T| \, dx\right]^2.$$

Here we used the Schwarz inequality in the form
$\int_0^1 f^2 \, du \geqq [\int_0^2 f \, du]^2$ with du corresponding to $T \, dx$.

By analogous manipulations,

$$M_t = -\int \nabla r \cdot C \cdot \nabla T \, dx \text{ and } |M_t| \leqq \int |\nabla r||C \cdot \nabla T| \, dx,$$

hence

$$|M_t| \leqq \int |C \cdot \nabla T| \, dx.$$

Combining inequalities,

(12) $$\qquad\qquad c_2 Q_t \geqq (M_t)^2.$$

This is a powerful inequality. Q is defined as it is in order to obtain (12).

The three inequalities

(10) $$\qquad\qquad Q \geqq \pm k + \tfrac{1}{2} n \log t$$

(11) $$\qquad\qquad M \geqq k e^{Q/n}$$

(12) $$\qquad\qquad c_2 Q_t \geqq (M_t)^2$$

and the qualitative fact $\lim M = 0$, as $t \to 0$, suffice by themselves to bound above and below both M and Q, as functions of time. No further reference to the differential equation is needed.

From $M(0) = 0$ and (12),

$$M \leqq \int_0^t (c_2 Q_t)^{\frac{1}{2}} dt,$$

whence

$$ke^{Q/n} \leqq M \leqq \int_0^t (c_2 Q_t)^{\frac{1}{2}} dt.$$

Now define $nR = Q \mp k - \frac{1}{2}n \log t$ in such a way that $R \geqq 0$ corresponds to (10). Then $Q_t = nR_t + n/2t$, and we obtain

$$kt^{\frac{1}{2}} e^R \leqq M \leqq (nc_2)^{\frac{1}{2}} \int_0^t (1/2t + R_t)^{\frac{1}{2}} dt.$$

When a and $a + b$ are positive $(a + b)^{\frac{1}{2}} \leqq a^{\frac{1}{2}} + b/2a^{\frac{1}{2}}$, hence

$$\int_0^t (1/2t + R_t)^{\frac{1}{2}} dt \leqq \int_0^t (1/2t)^{\frac{1}{2}} dt + \int_0^t (t/2)^{\frac{1}{2}} R_t \, dt$$

$$\leqq (2t)^{\frac{1}{2}} + R(t/2)^{\frac{1}{2}} - \int_0^t R/(8t)^{\frac{1}{2}} \, dt \leqq (2t)^{\frac{1}{2}} + R(t/2)^{\frac{1}{2}}$$

Here we used integration by parts and $R \geqq 0$ in the second and third steps. Applying this result,

$$kt^{\frac{1}{2}} e^R \leqq kM \leqq (2t)^{\frac{1}{2}} + R(t/2)^{\frac{1}{2}},$$

or

$$ke^R \leqq kM/t^{\frac{1}{2}} \leqq 2^{\frac{1}{2}}(1 + \frac{1}{2}R).$$

Clearly ke^R increases faster in R than $2^{\frac{1}{2}}(1 + \frac{1}{2}R)$ so that R must be bounded above. Therefore $M/t^{\frac{1}{2}}$ is bounded both above and below:

(13) $$kt^{\frac{1}{2}} \leqq M \leqq kt^{\frac{1}{2}}.$$

If we use best possible constants in (10) and (11), we can obtain

$$b_n(2c_1 nt)^{\frac{1}{2}} \leqq M \leqq (2c_2 nt)^{\frac{1}{2}} \left[1 + \min(\lambda, (\lambda/2)^{\frac{1}{2}}) \right],$$

where

$$b_n = (n/2t)^{\frac{1}{2}} \left\{ \pi^{\frac{1}{2}} / [\frac{1}{2}(n-1)]! \right\}^{1/n} \geqq 2^{-1/2n}$$

and

$$\lambda = \tfrac{1}{2}\log(c_2/c_1) - \log b_n \leqq (1/2n)\log 2 + \tfrac{1}{2}\log(c_2/c_1).$$

Thus λ is relatively small. Since $b_n \to 1$ as $n \to \infty$, the bounds sharpen with increasing n; indeed, they seem surprisingly sharp. For comparison, $M = (2nct)^{\frac{1}{2}}$ in the simple heat equation where $C_{ij} = c\delta_{ij}$ and $c_1 = c_2 = c$.

Part II: The G Bound

Here we obtain a result limiting the extent to which a fundamental solution can be very small over a large volume of space near its source point. From this result, we can show there is some overlap, defined as $\int \min(T_1, T_2)\, dx$, of two fundamental solutions with nearby source points, starting simultaneously.

Let T be $S(x, t, 0, 0)$ and let

$$(14) \qquad U(\xi, t) = t^{n/2} T\left(t^{\frac{1}{2}}\xi, t\right).$$

This coordinate transformation and renormalization makes $\int U d\xi = 1$, where $d\xi$ is the volume element. Furthermore, if μ is the constant such that $M \leqq \mu t^{\frac{1}{2}}$, we have $\int |\xi| U\, d\xi \leqq \mu$. For U, equation (1) transforms to

$$(15) \qquad 2tU_t = nU + \xi \cdot \nabla U + 2\nabla \cdot (C \cdot \nabla U).$$

Let

$$(16) \qquad G = \int \exp\left(-|\xi|^2\right) \log(U + \delta)\, d\xi,$$

where δ is a small positive constant. G is sensitive to areas where $|\xi|$ is not large and U is small. These tend to make G strongly negative. We later obtain a lower bound on G of the form

$$G \geqq -k(-\log\delta)^{\frac{1}{2}},$$

valid for sufficiently small δ. This bound limits the possibility for U to be small in a large portion of the region where $|\xi|$ is not large. From $U > 0$ the weak lower bound $G > \pi^{n/2}\log\delta$ follows immediately.

Differentiating (16) with respect to time and using (15), we obtain

$$2tG_t = H_1 + H_2 + H_3,$$

where

$$H_1 = n \int \exp\left(-|\xi|^2\right) U/(U + \delta)\, d\xi \gneqq 0,$$

$$H_2 = \int \exp\left(-|\xi|^2\right) \xi \cdot \nabla \log\left(U + \delta\right) d\xi$$

$$= -\int \nabla \cdot \left[\exp\left(-|\xi|^2\right) \xi\right] \log(U + \delta)\, d\xi,$$

by integration by parts, so that

$$H_2 = -\int \exp\left(-|\xi|^2\right) (\nabla \cdot \xi) \log(U + \delta)\, d\xi$$

$$+ \int \exp\left(-|\xi|^2\right) (2|\xi||\nabla|\xi|) \cdot \xi \log(U + \delta)\, d\xi$$

$$= -nG + 2 \int \exp\left(-|\xi|^2\right) |\xi|^2 \left[\log \delta + \log(1 + U/\delta)\right] d\xi,$$

hence,

$$H_2 \gneqq -nG + 2\log \delta \int |\xi|^2 \exp\left(-|\xi|^2\right) d\xi \gneqq -nG + n\pi^{n/2} \log \delta;$$

finally

$$H_3 = 2 \int \exp\left(-|\xi|^2\right) \nabla \cdot (C \cdot \nabla U)/(U + \delta)\, d\xi$$

$$= -2 \int \nabla \left[\exp\left(-|\xi|^2\right)/(U + \delta)\right] \cdot C \cdot \nabla U\, d\xi$$

$$= 4 \int \left(\exp\left(-|\xi|^2\right) \left[|\xi||\nabla|\xi| \cdot C \cdot \nabla U\right]/(U + \delta)\, d\xi$$

$$+ 2 \int \exp\left(-|\xi|^2\right) \left[\nabla U \cdot C \cdot \nabla U\right]/(U + \delta)^2\, d\xi$$

$$= H_3' + H_3'',$$

where

$$H_3' = 4 \int \exp\left(-|\xi|^2\right) \xi \cdot C \cdot \nabla \log(U + \delta)\, d\xi,$$

$$H_3'' = 2 \int \exp\left(-|\xi|^2\right) \nabla \log(U + \delta) \cdot C \cdot \nabla \log(U + \delta)\, d\xi.$$

From the Schwarz inequality,

$$(H_3')^2 \leqq \left\{ 4 \int \exp\left(-|\xi|^2\right) \xi \cdot C \cdot \xi \, d\xi \right\}$$

$$\times \left\{ 4 \int \exp\left(-|\xi|^2\right) \nabla \log(U + \delta) \cdot C \cdot \nabla \log(U + \delta) \, d\xi \right\}$$

$$\leqq \left\{ 4c_2 \int |\xi|^2 \exp\left(-|\xi|^2\right) d\xi \right\} 2H_3''$$

$$\leqq \left(4c_2 \cdot \tfrac{1}{2} n\pi^{\frac{1}{2}n} \right) \cdot 2H_3'' = 4nc_2\pi^{\frac{1}{2}n} H_3''.$$

Hence

$$\left| H_3' \right| \leqq k \left(H_3'' \right)^{\frac{1}{2}}.$$

Furthermore,

$$(17) \qquad H_3'' \geqq 2c_1 \int \exp\left(-|\xi|^2\right) |\nabla \log(U + \delta)|^2 \, d\xi.$$

Combining the lower bounds available for H_1, H_2, and H_3', we obtain

$$(18) \quad 2tG_t \geqq H_1 + H_2 + H_3'' - \left| H_3' \right|$$

$$\geqq 0 + \left(-nG + n\pi^{\frac{1}{2}n} \log \delta \right) + H_3'' - k \left(H_3'' \right)^{\frac{1}{2}}$$

$$\geqq k \log \delta - nG - k \left(H_3'' \right)^{\frac{1}{2}} + H_3''.$$

When we bound H_3'' below in terms of G, (18) will yield a lower bound on G.

A function $f(\xi) = f(\xi_1, \xi_2, \cdots, \xi_n)$ may be expanded in products of Hermite polynomials, of the form $\Pi H_{v(i)}(\xi_i)$, where the polynomials are defined and orthonormalized so that $\int_{-\infty}^{+\infty} \exp(-s^2) H_v(s) H_\lambda(s) \, ds = \delta_{v\lambda}$. The identity $dH_v(s)/ds = (2v)^{\frac{1}{2}} H_{v-1}(s)$ obtains, and the coefficients of these products in the similar expansion of, say, $\partial f/\partial \xi_i$ depend very simply on the coefficients in the expansion of f. If $\int \exp\left(-|\xi|^2\right) f \, d\xi = 0$, the coefficient of $\Pi H_0(\xi_i)$ is zero and we obtain

$$\int \exp\left(-|\xi|^2\right) |\nabla f|^2 \, d\xi = \sum \int \exp\left(-|\xi|^2\right) (\partial f/\partial \xi_i)^2 \, d\xi$$

$$\geqq 2 \int \exp\left(-|\xi|^2\right) f^2 \, d\xi.$$

Applying the above, with $f = \log(U + \delta) - \pi^{-n/2} G$, to (17), we obtain

(B19) $\quad H_3'' \geq 4c_1 \int \exp\left(-|\xi|^2\right) \left[\log(U + \delta) - \pi^{-n/2} G\right]^2 d\xi.$

The quantity $U^{-1}\left[\log(U + \delta) - \pi^{-n/2} G\right]^2$, related to the integrand in (19), is large for very small U, then decreases to zero, rises from zero to a local maximum at $U = U_c$, say, and finally decreases monotonically as $U \to \infty$ for $U \geq U_c$. (We know $\log \delta - \pi^{-n/2} G < 0$.) The equation for the maximum point U_c is $\log(U_c + \delta) - \pi^{-n/2} G = 2U_c/(U_c+\delta)$, from which $U_c < U_0 = \exp(2+\pi^{-n/2} G)$. Therefore the quantity under discussion is decreasing for $U \geq U_0$. The bound (8), $T \leq kt^{-n/2}$, corresponds to $U \leq k$. Hence the quantity has a lower bound of the form $k\left[\log(k + \delta) - kG\right]^2$ for $U \geq U_0$. Applying this to (19), we may say

$$H_3'' \geq 4c_1 \int \exp\left(-|\xi|^2\right) k\left[\log(k + \delta) - kG\right]^2 U^* d\xi,$$

where $U^* = U$ for $U > U_0$ and $U^* = 0$ for $U \leq U_0$. Thus we are ignoring the contribution to (19) of the region where $U \leq U_0$ and taking the worst case, U as large as possible, in the remaining region. For sufficiently negative G, the expression $\log(k + \delta) - kG$ will remain positive when δ is omitted, so that $\left[\log k - kG\right]^2 < \left[\log(k + \delta) - kG\right]^2$, and we can simplify the above inequality on H_3'' to the form

(20) $\quad\quad H_3'' \geq (k - kG)^2 \int \exp\left(-|\xi|^2\right) U^* d\xi.$

Let $\lambda = \int U^* d\xi$ and observe that $\int |\xi| U^* d\xi \leq \int |\xi| U d\xi \leq \mu$. Therefore

$$\mu \geq \int_{|\xi| \geq 2\mu/\lambda} |\xi| U^* d\xi \geq (2\mu/\lambda) \int_{|\xi| \geq 2\mu/\lambda} U^* d\xi,$$

hence,

$$\int_{|\xi| \geq 2\mu/\lambda} U^* d\xi \leq \tfrac{1}{2}\lambda, \quad \text{and so}$$

$$\int_{|\xi| \leq 2\mu/\lambda} U^* d\xi \geq \lambda - \tfrac{1}{2}\lambda = \tfrac{1}{2}\lambda.$$

This result can be applied to (20) and yields

(21) $$H_3'' \geqq (k - kG)^2 \cdot \exp\left(-(2\mu/\lambda)^2\right) \left(\tfrac{1}{2}\lambda\right).$$

This is not effective unless we can bound λ below, or bound $\int \hat{U}\, d\xi = 1 - \lambda$ above, where $\hat{U} = U - U^*$ so that $\hat{U} = 0$ unless $U \leqq U_0$, in which case $\hat{U} = U$. Of course, we know $\int |\xi|\hat{U}\, d\xi \leqq \mu$ because $\hat{U} \leqq U$. Under the moment constraint and the constraint $\hat{U} \leqq U_0$, the maximum of $\int \hat{U}\, d\xi$ is clearly realized by having $\hat{U} = U_0$ for $|\xi| \leqq \rho$ and $\hat{U} = 0$ for $|\xi| > \rho$, where ρ is such that

$$\int |\xi|\hat{U}\, d\xi = \int |\xi|U_0\, d\xi = \left[n\pi^{n/2}/(n+1)(n/2)!\right] \rho^{n+1} U_0 = \mu.$$

This makes

$$1 - \lambda = \int \hat{U}\, d\xi = \left[\pi^{n/2}/(n/2)!\right]\rho^n U_0,$$

$$1 - \lambda \leqq U_0(k\mu/U_0)^{n/(n+1)} \quad \text{or} \quad 1 - \lambda \leqq kU_0^{1/(n+1)}.$$

If U_0, which is $\exp(2 + \pi^{-n/2}G)$, is small enough, then $1 - \lambda$ is small and λ is bounded below. Thus $\lambda \geqq \tfrac{1}{2}$, say, for all sufficiently large $-G$. Now from (21), we have

$$H_3'' \geqq (k - kG)^2$$

for sufficiently large $-G$.

Returning to inequality (18) controlling G_t and applying the above result, we can state that *for sufficiently negative* G,

(22)
$$2dG/d(\log t) = 2tG_t \geqq -nG + k\log \delta + (k - kG)^2$$
$$- k(k - kG)$$
$$\geqq k|G|^2 + k\log \delta.$$

Let $G_1(c_1, c_2, n)$ be the number such that when $G \leqq G_1$, we know G is small enough to make (22) valid. Let $G_2(c_1, c_2, n, \delta) = -k(-\log \delta)^{\frac{1}{2}}$ be the largest number such that $k|G|^2 + k\log \delta > 0$ for all $G < G_2$. Then $\min(G_1, G_2) = G_3$ is the smallest possible value of G. If we had $G(t_1) = G_3 - \epsilon$, we would have $dG/d(\log t) \geqq \epsilon^*$ for all $t \leqq t_1$, and consequently $G(t) \leqq G(t_1) - \epsilon^* \log(t_1/t)$, which implies $G \to -\infty$ as $t \to 0$. But since $G \geqq \pi^{n/2}\log \delta$, the hypothesis $G(t_1) = G_3 - \epsilon$ is impossible. Our conclusion is $G \geqq G_3$, or

J O H N   F.   N A S H ,   J R .

Continuity of Solutions of Parabolic and Elliptic Equations

(23)
$$G \geqq -k(-\log \delta)^{\frac{1}{2}}$$

for all *sufficiently small values of* δ, because $G_2 \leqq G_1$ and

$$G_3 = G_2 = -k(-\log \delta)^{\frac{1}{2}}$$

when δ is small enough.

Part III: The Overlap Estimate

Let T_1 and T_2 be two fundamental solutions $S(x, t, x_1, 0)$ and $S(x, t, x_2, 0)$ with nearby sources. Change coordinates, defining $U_1 = t^{n/2} T_1(t^{\frac{1}{2}}\xi, t)$ and $U_2 = t^{n/2} T_2(t^{\frac{1}{2}}\xi, t)$. Let $\xi_1 = x_1/t^{\frac{1}{2}}$ and $\xi_2 = x_2/t^{\frac{1}{2}}$. Here the source of the (renormalized) fundamental solution U_i is ξ_i rather than the origin, which was the source of U in Part II. Taking this into account, we apply (23), obtaining

$$\int \exp(-|\xi - \xi_i|^2) \log(U_i + \delta) \, d\xi = G_i \geqq -k(-\log \delta)^{\frac{1}{2}},$$

where $i = 1$ or 2 and δ must be sufficiently small. We may add the inequalities above and obtain

$$\int \max_i \left[\exp\left(-|\xi - \xi_i|^2\right)\right] \max_i \left[\log(U_i + \delta)\right] d\xi +$$

$$\int \min_i \left[\exp(-|\xi - \xi_i|^2)\right] \min_i \left[\log(U_i + \delta)\right] d\xi \geqq -2k(-\log \delta)^{\frac{1}{2}}$$

in which we form two integrals with sum at least as large as the sum of the original integrals. We abbreviate the above to

$$\int f^* \log(U_{\max} + \delta) \, d\xi + \int \hat{f} \log(U_{\min} + \delta) \, d\xi \geqq -k(-\log \delta)^{\frac{1}{2}}.$$

For the first integral, we observe (assuming $\delta \leqq 1$)

$$\int f^* \log(U_{\max} + \delta) \, d\xi \leqq \int f^*(U_1 + U_2) \, d\xi$$

$$\leqq \int (U_1 + U_2) \, d\xi = 2.$$

For the second integral,

$$\int \hat{f} \log(U_{\min} + \delta) \, d\xi \leqq \log \delta \int \hat{f} \, d\xi + \max \left[\hat{f}\right] \int \log(1 + U_{\min}/\delta) \, d\xi$$

$$\leqq w \log \delta + \delta^{-1} \int U_{\min} \, d\xi,$$

where

$$w = \int \min \left[\exp\left(-|\xi - \xi_1|^2\right), \exp\left(-|\xi - \xi_2|^2\right) \right] d\xi.$$

Therefore we obtain

$$2 + w \log \delta + \delta^{-1} \int \min(U_1, U_2) \, d\xi \geqq -k(-\log \delta)^{\frac{1}{2}},$$

or

$$(24) \qquad \int \min(T_1, T_2) \, dx = \int \min(U_1, U_2) \, d\xi$$
$$\geqq \delta \left[-2 - w \log \delta - k(-\log \delta)^{\frac{1}{2}} \right].$$

This is valid for sufficiently small δ, say for $\delta \leqq \delta_1$. Also, there is a value $\delta_2(w)$ such that for $\delta < \delta_2(w)$, the bracketed expression is positive. If we set $\delta = \frac{1}{2} \min(\delta_1, \delta_2)$, the right member of (24) is definitely positive, and we may conclude

$$(25) \qquad \int \min(T_1, T_2) \, dx \geqq \phi\left(|\xi_1 - \xi_2|\right) \geqq \phi(|x_1 - x_2|)/t^{\frac{1}{2}})$$

because w is a function of $|\xi_1 - \xi_2|$. The function ϕ is decreasing but always positive. It is an a priori function, determined only by c_1, c_2, and n. This inequality (25) is our first estimate on the overlap of fundamental solutions. Its weakness is that we know little about the function ϕ.

Part IV: Continuity in Space

We can obtain a strong inequality by iterative use of (25). Observe that

$$(26) \qquad \frac{1}{2} \int |T_1 - T_2| \, dx = \frac{1}{2} \int [T_1 + T_2 - 2\min(T_1, T_2)] \, dx$$
$$\leqq 1 = \phi\left(|x_1 - x_2|/t^{\frac{1}{2}}\right) = \psi\left(|x_1 - x_2|/t^{\frac{1}{2}}\right)$$

in which we define the function ψ, which is increasing but always less than one.

Let $T_a = \max(T_1 - T_2, 0)$ and $T_b = \max(T_2 - T_1, 0)$ so that

$T_a + T_b = |T_1 - T_2|$ and $\int (T_a - T_b)\, dx = \int (T_1 - T_2)\, dx = 0$.
Then

$$\int T_a\, dx = \int T_b\, dx = A(t) = \tfrac{1}{2} \int |T_1 - T_2|\, dx \leqq \psi (|x_1 - x_2|/t^{\frac{1}{2}}),$$

defining $A(t)$. Let

$$\chi (x, \bar{x}, t) = T_a(x)\, T_b(\bar{x})/A(t).$$

Let $T_a^*(x', t', t)$ be the bounded solution in x' and t' of (1) defined for
$t' \geqq t$ and having the initial value $T_a^*(x, t, t) = T_a(x, t)$. Define T_b^*
similarly. Then from (4),

$$T_a^*(x', t', t) = \int S(x', t', x, t)\, T_a(x, t)\, dx$$

$$= \int\!\!\int S(x', t', x, t)\chi (x, \bar{x}, t)\, dx\, d\bar{x},$$

and

$$T_1(x', t') - T_2(x', t') = T_a^* - T_b^*$$

$$= \int\!\!\int [(x', t', x, t) - S(x', t', \bar{x}, t)]\, \chi (x, \bar{x}, t)\, dx\, d\bar{x}$$

by the superposition principle ($T_1 - T_2$ and $T_a^* - T_b^*$ are both solutions
of (1) for $t' \geqq t$, and by definition, $T_a^* - T_b^* = T_1 - T_2$ at $t' = t$.).
Integrating this over dx', we obtain

$$\tfrac{1}{2} \int |T_1(x', t') - T_2(x', t')|\, dx'$$

$$\leqq \int\!\!\int\!\!\int |S(x', t', x, t) - S(x', t', \bar{x}, t)|\, dx'\, \chi (x, \bar{x}, t)\, dx\, d\bar{x},$$

whence

(27) $\qquad A(t') \leqq \int\!\int \psi (|x - \bar{x}|/(t' - t)^{\frac{1}{2}})\chi (x, \bar{x}, t)\, dx\, d\bar{x}$

by application of (26). Incidentally, the right member above is

$$\leqq \int\!\!\int \chi (x, \bar{x}, t)\, dx\, d\bar{x} = A(t);$$

thus $A(t') \leqq A(t)$ when $t' \geqq t$. This inequality (27) is the key to the
iterative argument which strengthens (25) and (26).

　　To begin the iterative argument, we choose any specific number

d and let $\epsilon = \phi(d) = 1 - \psi(d)$. (If we were trying to get an explicit formula for the exponent α in (2), we would choose d with regard to an explicit formula for $\phi(d)$ so as to optimize the result.) Let $\sigma = 1 - \epsilon/4$. For each integer ν, let t_ν be the time (or the least time) at which $A(t) = A(t_\nu) = \sigma^\nu$, if t_ν exists. This is in reference to a specific pair, T_1 and T_2, of fundamental solutions. We know, for example, that $t_1 < \tau$, where $\tau = |x_1 - x_2|^2/d^2$, because $A(r) \leqq \psi(|x_1 - x_2|/\tau^{\frac{1}{2}}) = \psi(d) = 1 - \epsilon$ and $\sigma = 1 - \epsilon/4 > 1 - \epsilon$, so that $A(\tau) < A(t_1) = \sigma$.

Let $M_a(t) = \int |x - x_0| T_a\, dx$, where x_0 is $\frac{1}{2}(x_1 + x_2)$, the midpoint of the line segment joining the source points x_1 and x_2 of the fundamental solutions T_1 and T_2. Define M_b similarly and let $M_\nu = \max [M_a(t_\nu), M_b(t_\nu)]$. We decompose T_a into nearer and farther parts T_a' and $T_a - T_a'$ at each time t_ν as follows: for $|x - x_0| \leqq 2\sigma^{-\nu} M_\nu$, define $T_a' = T_a$; otherwise $T_a' = 0$. Then $2\sigma^{-\nu} M_\nu \int (T_a - T_a')\, dx \leqq \int |x - x_0|(T_a - T_a')\, dx \leqq \int |x - x_0| T_a\, dx \leqq M_\nu$, and consequently, $\int (T_a - T_a')\, dx \leqq \frac{1}{2}\sigma^\nu$ and $\int T_a'\, dx \geqq \frac{1}{2}\sigma^\nu$. Define T_b' similarly and define $\chi_\nu'(x, \bar{x}) = \sigma^{-\nu} T_a'(x) T_b'(\bar{x})$. Now, applying (27) with $t = t_\nu$, we can say

$$A(t') \leqq \iint \psi(|x - \bar{x}|)/(t' - t_\nu)^{\frac{1}{2}} \Big[\{\chi(x, \bar{x}, t_\nu) - \chi_\nu'(x, \bar{x})\} $$
$$+ \chi_\nu'(x, \bar{x})\Big]\, dx\, d\bar{x}$$

$$\leqq \iint \{\chi - \chi_\nu'\}\, dx\, d\bar{x} + \psi\left(4\sigma^{-\nu} M_\nu/(t' - t_\nu)^{\frac{1}{2}}\right)$$
$$\iint \chi_\nu'\, dx\, d\bar{x},$$

because when $\chi_\nu' > 0$, we know both $T_a' > 0$ and $T_b' > 0$ so that both $|x - x_0|$ and $|\bar{x} - x_0|$ are $\leqq 2\sigma^{-\nu} M_\nu$, and consequently, $|x - \bar{x}| \leqq 4\sigma^{-\nu} M_\nu$, and we also know that $\chi \geqq \chi_\nu'$ and $\psi < 1$. Proceeding further,

$$A(t') \leqq \iint \chi\, dx\, d\bar{x} - \left[1 - \psi\left(4\sigma^{-\nu} M_\nu/(t' - t_\nu)^{\frac{1}{2}}\right)\right]$$
$$\iint \chi_\nu'\, dx\, d\bar{x}$$

$$\leqq \sigma^\nu - [1 - \psi]\sigma^{-\nu} \int T_a'\, dx \int T_b'\, dx$$

$$\leq \sigma^v - [1 - \psi]\sigma^{-v}(\sigma^v/2)^2$$

$$\leq \sigma^v \left[3/4 + 1/4\psi \left(4\sigma^{-v} M_v/(t' - t_v)^{\frac{1}{2}}\right)\right].$$

We now set $t' = t_v + 16\sigma^{-2v}(M_v)^2 d^{-2}$, and the argument of ψ above becomes d. Then since $\psi(d) = 1 - \epsilon$, we obtain

$$A(t') \leq \sigma^v \left[3/4 + 1/4(1 - \epsilon)\right] = \sigma^v(1 - \epsilon/4) = \sigma^{v+1}.$$

Hence

230 (28) $$t_{v+1} \leq t' = t_v + 16\sigma^{-2v}(M_v)^2 d^{-2}.$$

This will bound the sequence $\{t_v\}$ of times after we obtain a bound on the sequence $\{M_v\}$ of moments.

Observe that

$$T_a(x', t') = \max(T_1(x', t') - T_2(x', t'), 0)$$
$$= \max(T_a^*(x', t', t) - T_b^*(x', t', t), 0) \leq T_a^*(x', t', t)$$
$$= \int S(x', t', x, t) T_a(x, t)\, dx.$$

Therefore

$$M_a(t') = \int |x' - x_0| T_a(x', t')\, dx'$$
$$\leq \int\int \left[|x' - x| + |x - x_0|\right] S(x', t', x, t) T_a(x, t)\, dx\, dx';$$

hence

$$M_a(t') \leq \int |x - x_0| T_a(x, t) \int S(x', t', x, t)\, dx'\, dx$$
$$+ \int T_a(x, t) \int |x' - x| S(x', t', x, t)\, dx'\, dx,$$

or

$$M_a(t') \leq \int |x - x_0| T_a(x, t)\, dx + \mu(t' - t)^{\frac{1}{2}} \int T_a(x, t)\, dx$$
$$\leq M_a(t) + A(t)\mu(t' - t)^{\frac{1}{2}}.$$

Now let t and t' be t_v and t_{v+1}, use a similar estimate for M_b and the definition $M_v = \max(M_a(t_v), M_b(t_v))$, and obtain, by (28),

$$M_{v+1} \leq M_v + \sigma^v \mu(t_{v+1} - t_v)^{\frac{1}{2}}$$

$$\leq M_v + \sigma^v \mu \left(16\sigma^{-2v}(M_v)^2 d^{-2}\right)^{\frac{1}{2}} \leq M_v(1 + 4\mu/d).$$

Now $t_0 = 0$ and $M_0 = M_a(t_0) = M_b(t_0) = \frac{1}{2}|x_1 - x_2|$, because T_1 and T_2 concentrate at x_1 and x_2 as $t \to 0$, and $|x_1 - x_0| = |x_2 - x_0| = \frac{1}{2}|x_1 - x_2|$ since $x_0 = \frac{1}{2}(x_1 + x_2)$. Therefore we have

$$M_v \leq \tfrac{1}{2}|x_1 - x_2|(1 + 4\mu/d)^v.$$

With this and (28), the sequence $\{t_v\}$ can be bounded:

$$t_{v+1} \leq t_v + 16\sigma^{-2v}\left[\tfrac{1}{2}|x_1 - x_2|(1 + 4\mu/d)^v\right]^2 d^{-2}.$$

Hence

$$t_{v+1} \leq 4d^{-2}|x_1 - x_2|^2 \sum_{\lambda=0}^{v} \left[(1 + 4\mu/d)/\sigma)\right]^{2\lambda}.$$

Summing this geometrical series,

$$t_v/|x_1 - x_2|^2 \leq 4d^{-2}\left\{ \frac{(\sigma^{-2}(1 + 4\mu/d))^{2v+2}}{\left[\sigma^{-2}(1 + 4\mu/d) - 1\right]} \right\} \equiv \xi\eta^v$$

(definition of ζ, η). Now for any time t, define $v(t)$ to be either zero or the integer such that

$$\zeta\eta^{v(t)} \leq t/|x_1 - x_2|^2 < \zeta\eta^{v(t)+1}$$

if this integer exists. Then $t_{v(t)} \leq t$ and $A(t) \leq A(t_{v(t)}) = \sigma^{v(t)}$. Also,

$$v(t) \geq \left(\log\left(t/\zeta|x_1 - x_2|^2\right)/\log\eta\right) - 1.$$

From these observations, we conclude

$$\sigma^{v(t)} \leq \sigma^{-1} \exp\left[(\log\sigma/\log\eta)\log\left(t/\zeta|x_1 - x_2|^2\right)\right];$$

hence

$$A(t) \leq \sigma^{-1}\left(t/\zeta|x_1 - x_2|^2\right)^{\log\sigma/\log\eta},$$

or

$$\tfrac{1}{2}\int |T_1 - T_2|\, dx \leq \sigma^{-1}\zeta^{\alpha/2}\left(|x_1 - x_2|/t^{\frac{1}{2}}\right)^\alpha,$$

where $\frac{1}{2}\alpha = -\log\sigma/\log\eta$.

Both σ and η are determined by d. Specifically, $\sigma = 1 - \frac{1}{4}\phi(d)$

and $\eta = \left[\sigma^{-2}(1 + 4\mu/d)\right]^2$. An optimal choice of d in relation to $\phi(d)$ would maximize α. We may choose d arbitrarily as, $d^2 = c_1$, say; this will make α a function of μ and c_2/c_1 (proof omitted). In any case, even if we set $d = 1$, we obtain the estimate

$$(29) \int |S(x, t, x_1, t_0) - S(x, t, x_2, t_0)| \leq A_1 \left(|x_1 - x_2|/(t - t_0)^{\frac{1}{2}}\right)^\alpha,$$

where A_1 and α are a priori constants depending only on n, c_1 and c_2. Also, for the dual adjoint equation,

$$(30) \qquad \int |S(x_1, t, x_0, t_0) - S(x_2, t, x_0, t_0)| \, dx_0$$
$$\leq A_1 \left(|x_1 - x_2|/(t - t_0)^{\frac{1}{2}}\right)^\alpha.$$

With (30), we obtain the estimate for the continuity in space of a bounded solution of (1). If $T(x, t)$ satisfies (1) and $|T| \leq B$ for $t \geq t_0$, then

$$|T(x_1, t) - T(x_2, t)| \leq \left|\int [S(x_1, t, x_0, t) - S(x_2, t, x_0, t_0)]\right.$$
$$\left. T(x_0, t_0) \, dx_0 \right|$$
$$\leq B \int |S(x_1, t, x_0, t_0) - S(x_2, t, x_0, t_0)| \, dx_0.$$

Hence,

$$(31) \qquad |T(x_1, t) - T(x_2, t)| \leq BA_1 \left(|x_1 - x_2|/(t - t_0)^{\frac{1}{2}}\right)^\alpha.$$

Part V: Time Continuity

Equation (31) gives half of (2); the remaining part, time continuity, can be derived from (31) and the moment bound (13). Let $T(x, t)$ be a solution of (1) with $|T| \leq B$ for $t \geq t_0$. Then for $t' > t > t_0$ we have

$$T(x, t) - T(x, t') = T(x, t) - \int S(x, t', \bar{x}, t) T(\bar{x}, t) \, d\bar{x}$$
$$= \int S(x, t', \bar{x}, t) [T(x, t) - T(\bar{x}, t)] \, d\bar{x},$$

since $\int S \, d\bar{x} = 1$. Therefore,

$$|T(x, t) - T(x, t')| \leqq \int S(x, t', \tilde{x}, t) | T(x, t) - T(\tilde{x}, t) | \, d\tilde{x}$$

$$\leqq \int S(x, t', x + y, t) | T(x, t) - T(x + y, t) | \, dy.$$

Now we separate this integral into two parts, in terms of a radius ρ; one where $|y| \leqq \rho$ and one where $|y| > \rho$. Thus $|T(x, t) - T(x, t')| \leqq I_1 + I_2$, where

$$I_1 = \int_{|y| \leqq \rho} S(x, t', x + y, t) | T(x, t) - T(x + y, t) | \, dy$$

$$\leqq B A_1 \left(\rho / (t - t_0)^{\frac{1}{2}} \right)^{\alpha}$$

(because $\int S \, dy = 1$), and

$$I_2 = \int_{|y| > \rho} S(x, t', x + y, t) \left| T(x, t) - T(x + y, t) \right| \, dy$$

$$\leqq 2B\rho^{-1} \int_{|y| > \rho} |y| \, S(x, t', x + y, t) \, dy \leqq 2B\mu (t' - t)^{\frac{1}{2}} / \rho.$$

Adding the two inequalities,

$$\left| T(x, t) - T(x, t') \right| \leqq B A_1 \left(\rho / (t - t_0)^{\frac{1}{2}} \right)^{\alpha} + 2B\mu (t' - t)^{\frac{1}{2}} / \rho,$$

and if we choose ρ so as to minimize the sum, then

$$\alpha A_1 \rho^{1+\alpha} = 2\mu (t' - t)^{\frac{1}{2}} (t - t_0)^{\frac{1}{2}\alpha},$$

and we obtain

(32) $\quad |T(x, t) - T(x, t')| \leqq B A_2 \left[(t' - t) / (t - t_0) \right]^{\frac{1}{2}\alpha / (1+\alpha)},$

where $A_2 = (1 + \alpha) A_1 (2\mu / \alpha A_1)^{\alpha / (1+\alpha)}$. This result (32), combined with (31) yields (2), with $A = \max(A_1, A_2)$.

Part VI: Elliptic Problems

We treat elliptic problems as a special type of parabolic problem, one in which the coefficients of the equation are time independent and a time independent solution is sought. The Hölder continuity of solutions of uniformly elliptic equations of the form $\nabla \cdot (C \cdot \nabla T) = 0$ appears as a corollary of the result for the parabolic case. There may exist another proof of our result (3). P. R. Garabedian writes from London of a manuscript by Ennio de Giorgi containing such a result. See de Giorgi's note [9].

Let $\mathfrak{D}$ be a domain in space-time defined by the constraints $|x| \leq \sigma$ and $t \geq 0$. Then $\mathfrak{D}$ is a solid semi-infinite spherical cylinder. Call $\mathfrak{B}$ the points of the cylindrical surface or boundary of $\mathfrak{D}$, where $|x| = \sigma$. Let $\mathfrak{D}_0$ be the points of the base of $\mathfrak{D}$, where $t = 0$. Define $\mathfrak{B}^*$ as the total boundary of $\mathfrak{D}$, the union $\mathfrak{B} \cup \mathfrak{D}_0$, of the base and cylindrical surfaces.

A "Dirichlet parabolic boundary value problem" is given when values of T are specified on $\mathfrak{B}^*$ and we ask for a solution of (1) in $\mathfrak{D}$ assuming these specified values on $\mathfrak{B}$. The solution of the problem must depend linearly on the boundary values; also, the maximum and minimum principles must hold. These facts require that the solution $T(x, t)$ be determined in this way:

$$(33) \qquad T(x, t) = \int T(\xi)\, d\rho(\xi; x, t).$$

Here (x, t) is a point of $\mathfrak{D}$, ξ is any point of $\mathfrak{B}^*$, and $d\rho(\xi; x, t)$ is a positive measure, associated with ξ, which has $\int d\rho = 1$ and which vanishes for $t(\xi) > t$. The time and space coordinates of the point ξ are called $t(\xi)$ and $x(\xi)$. We cannot pause here for a detailed justification of (33), but refer the reader to the literature.

We can define a boundary value problem for which we know the solution in advance by setting $T(\xi) = S(x(\xi), t(\xi), x_0, t_0)$ if $t_0 < 0$. Then the solution of the problem is $S(x, t, x_0, t_0)$, and from (33),

$$(34) \qquad S(x, t, x_0, t_0) = \int S(x(\xi), t(\xi), x_0, t_0)\, d\rho(\xi; x, t).$$

This is a powerful identity; it enables us to convert information on fundamental solutions into information on $d\rho$, and in particular, we can obtain a moment bound for $d\rho$. Multiplying (34) by $|x - x_0|$ and integrating, we have

$$\int |x - x_0| S(x, t, x_0, t_0)\, dx_0$$
$$= \int\int |x - x_0| S(x(\xi), t(\xi), x_0, t_0)\, d\rho\, dx_0.$$

Hence

$$\mu(t - t_0)^{\frac{1}{2}}$$
$$\geqq \iint \{|x - x(\xi)| - |x_0 - x(\xi)|\} S(x(\xi), t(\xi), x_0, t) \, d\rho \, dx_0.$$

so that

$$\mu(t - t_0)^{\frac{1}{2}} + \iint |x_0 - x(\xi)| S(x(\xi), t(\xi), x_0, t_0) \, dx_0 \, d\rho$$
$$\geqq \int |x - x(\xi)| \int S(x(\xi), t(\xi), x_0, t_0) \, dx_0 \, d\rho.$$

Since $\int S \, dx_0 = 1$, and from the moment bound (13) again, we obtain

$$\mu(t - t_0)^{\frac{1}{2}} + \int \mu(t(\xi) - t_0)^{\frac{1}{2}} \, d\rho \geqq \int |x - x(\xi)| \, d\rho.$$

Now $d\rho$ vanishes unless $t(\xi) \leqq t$, and t_0 can be as near to zero as desired: also, $\int d\rho = 1$. Hence we can simplify the above to:

(35)
$$2\mu t^{\frac{1}{2}} \geqq \int |x - x(\xi)| \, d\rho(\xi; x, t).$$

This moment bound (35) on $d\rho$ enables us to control the relative sizes of the effects of the two parts of the boundary in determining $T(x, t)$, where (x, t) is in $\mathfrak{D}$. Thus

$$\int |x - x(\xi)| \, d\rho \geqq \int ||x| - |x(\xi)|| \, d\rho \geqq (\sigma - |x|) \int_{\mathfrak{B}} d\rho.$$

Hence

(36)
$$\int_{\mathfrak{B}} d\rho(\xi; x, t) \leqq 2\mu t^{\frac{1}{2}} / (\sigma - |x|).$$

Now let $T(x)$ be a solution in a region $\mathfrak{R}$ of n-space of $\nabla \cdot (C(x) \cdot \nabla T) = 0$, where $C(x)$ satisfies the uniform ellipticity condition with bounds c_1 and c_2. If we introduce time and define $T(x, t) = T(x)$, then $T(x, t)$ satisfies $\nabla \cdot (C \cdot \nabla T) = T_t$, which is of our form (1). Suppose x_1 and x_2 are two points of $\mathfrak{R}$ and let $d(x_1, x_2)$ be the smaller of $d(x_1)$ and $d(x_2)$, the distances from the boundary of $\mathfrak{R}$ of x_1 and x_2 (of course, $d(x_1, x_2)$ may be infinite). For any $\sigma < d(x_1, x_2)$, we can define $\mathfrak{D}_1$ as the set of points (x, t) in space-time where $|x - x_1| \leqq \sigma$ and $t \geqq 0$; also, $\mathfrak{D}_2$ can be defined for x_2 and the boundaries $\mathfrak{B}_1, \mathfrak{B}_2,$

etc. can be defined in the obvious way. $T(x, t)$ can be regarded as a solution of a paraboic boundary value problem either in $\mathcal{D}_1$ or $\mathcal{D}_2$. Another problem with solution $T'(x, t)$ can be defined at first as an initial value problem in all space by setting $T'(x, 0) = T(x)$ for all x where $\min(|x - x_1|, |x - x_2|) \leq \sigma$, that is, $T'(x, t) = T(x)$ when $(x, t) \in \mathcal{D}_{10} \cup \mathcal{D}_{20}$, and setting $T'(x, 0) = 0$ for all other x values. If $B(\sigma) = \max |T(x)|$ over the set of x values where $\min(|x - x_1|, |x - x_2|) \leq \sigma$, then $|T'(x, 0)| \leq B(\sigma)$; furthermore, the solution $T'(x, t)$ satisfies $|T'| \leq B(\sigma)$ for all $t \geq 0$ by the maximum principle. We can also regard $T'(x, t)$ as a solution of a boundary value problem, either in $\mathcal{D}_1$ or in $\mathcal{D}_2$, where the boundary values are just the values $T'(x(\xi), t(\xi))$ assumed there anyway.

By (33), for any $(x, t) \in \mathcal{D}_i$,

$$T(x, t) - T'(x, t) = \int \left[T(x(\xi), t(\xi)) - T'(x(\xi), t(\xi)) \right] d\rho_i(\xi; x, t),$$

where $d\rho_i$ is the measure associated with $\mathcal{D}_i$ and $i = 1, 2$. Now $T(x, t) = T(x)$ is time independent, and on $\mathcal{D}_{i0}$ we have $T(x, t) = T'(x, t) = T(x)$. Therefore,

$$\left| T(x) - T'(x, t) \right| \leq \int_{\mathcal{B}_i} \left| T(x(\xi)) - T'(x(\xi), t(\xi)) \right| d\rho_i$$

$$\leq 2B(\sigma) \int_{\mathcal{B}_i} d\rho_i,$$

and

$$\left| T(x_i) - T'(x_i, t) \right| \leq 4B(\sigma)\mu t^{\frac{1}{2}}/\sigma,$$

by use of (36). With our Hölder continuity estimate (2) for solutions of $\nabla \cdot (C \cdot \nabla T) = T_t$ in free space, we can bound $|T'(x_1, t) - T'(x_2, t)|$. This, with the inequality above yields

$$|T(x_1) - T(x_2)| \leq B(\sigma)A\left(|x_1 - x_2|/t^{\frac{1}{2}} \right)^{\alpha} + 8\mu B(\sigma)t^{\frac{1}{2}}/\sigma,$$

valid for any positive t. Choice of the optimal t value gives an inequality of the form

(37) $\qquad |T(x_1) - T(x_2)| \leq B(\sigma)A'(|x_1 - x_2|/\sigma)^{\alpha/(\alpha+1)}.$

If $|T(x)| \leq B$ in $\mathfrak{R}$, we may set $\sigma = d(x_1, x_2)$ and obtain (3).

Appendix

The methods used above can give more explicit results, such as an explicit lower bound for the Hölder exponent α. This takes the form $\alpha = \exp\left[-a_n(\mu^2/c_1)^{n+1}\right]$, where a_n depends only on the dimension n. However, a sharper estimate for α might take a quite different form. Numerical calculation of extremal examples might give a better picture.

The moment bound (13) serves to control the rate of dispersal of fundamental solutions. An iterative argument based on (33) and (35) obtains stronger results from (13). In this argument, a fundamental solution is treated as the solution of an array of parabolic boundary value problems, the boundaries being a sequence of spheres centered at the source of the fundamental solution. The result is as follows: let $\nu = [\rho/2\mu(t_2 - t_1)^{\frac{1}{2}}]$, the largest integer not greater than $\rho/2\mu(t_2 - t_1)^{\frac{1}{2}}$, then

$$\int_{|x_2-x_1| \geq \rho} S(x_2, t_2, x_1, t_1)\, dx_2 \leq (\pi/4)^{\nu/2}/(\nu/2)!$$

$$\leq \exp\left[-\tfrac{1}{2}(\nu + 1)\log(2(\nu + 1)\pi e)\right].$$

Hence,

$$(38) \quad \int_{|x_2-x_1| \geq \rho} S(x_2, t_2, x_1, t_1)\, dx_2$$

$$\leq \exp\left\{-\rho \log\left(\rho/\pi e\mu(t_2 - t_1)^{\frac{1}{2}}\right)/4\mu(t_2 - t_1)^{\frac{1}{2}}\right\}.$$

With (38), the reproductive identity (5), and the bound (7), we obtain a pointwise upper bound of the form

$$S(x_2, t_2, x_1, t_1) \leq k(t_2 - t_1)^{-n/2}$$

$$(39) \qquad \exp\left[-k|x_1 - x_2|(t_2 - t_1)^{-\frac{1}{2}}\right.$$

$$\left. \log\left(k|x_1 - x_2|(t_2 - t_1)^{-\frac{1}{2}}\right)\right].$$

On the other hand, we obtain from (5) and (23) (or alternatively, from (38) and an analogue of (25)), by an argument resembling that which gave (25), the lower bound

$$(40) \quad S(x_2, t_2, x_1, t_1) \geq (t_2 - t_1)^{-n/2}\phi^*\left(|x_1 - x_2|/(t_2 - t_1)^{\frac{1}{2}}\right),$$

JOHN F. NASH, JR.

Continuity of Solutions of Parabolic and Elliptic Equations

where ϕ^* is an a priori function determined by c_1, c_2, and n. The inequality $S(x_2, t_2, x_1, t_1) \geqq P_a P_b P_c$, where

$$P_a = \min S\left(x_2, t_2, \bar{x}, \tfrac{1}{2}(t_1 + t_2)\right) \quad \text{for} \quad |\bar{x} - x_1| \leqq \rho,$$
$$P_b = \min S\left(\bar{x}, \tfrac{1}{2}(t_1 + t_2), x_1, t_1\right) \quad \text{for} \quad |\bar{x} - x_2| \leqq \rho,$$
$$P_c = \int d\bar{x}, \quad \text{where} \quad |\bar{x} - x_1| \leqq \rho \quad \text{and} \quad |\bar{x} - x_2| \leqq \rho,$$

can be used in a iterative argument to strengthen (40). For any $\epsilon > 0$, we obtain

(41)
$$S(x_2, t_2, x_1, t_1) \geqq k_1 (t_2 - t_1)^{-n/2} \exp\left[-k_2 \left(|x_1 - x_2|/(t_2 - t_1)^{\frac{1}{2}} \right)^{2+\epsilon} \right],$$

where k_1 and k_2 depends on ϵ (and on c_1, c_2, and n).

With (38), (41) and (35), we can estimate the speed of convergence to assigned boundary values of the solution of an elliptic boundary value problem, provided the boundary is "tame" enough. A point ξ on the boundary $\mathfrak{B}$ of a region $\mathfrak{R}$ is called *regular* if there are two positive numbers ρ and ϵ such that any sphere with radius $\leqq \rho$ and centered at ξ has at least the fraction ϵ of its volume *not* within $\mathfrak{B}$. Then there are constants $\mathfrak{D}, \sigma$, and β determined by $\epsilon, c_2/c_1$, and n such that for any x in $\mathfrak{R}$ with $|x - \xi| \leqq \sigma\rho$, we have

(42)
$$T(x) \geqq \min T(\bar{\xi}) - D|(x - \xi)/\rho|^\beta,$$
$$T(x) \leqq \max T(\bar{\xi}) + D|(x - \xi)/\rho|^\beta, \quad \text{where} \quad |\bar{\xi} - \xi| \leqq \rho$$

($\bar{\xi}$ represents a variable point on the boundary $\mathfrak{B}$).

From (42), it follows that the solution of an elliptic boundary value problem is continuous at the boundary if continuous values were assigned on the boundary and all boundary points are regular. With Hölder continuous boundary values, the solution is Hölder continuous in the region and at the boundary.

From the estimates above, we can fairly easily derive a "Harnack inequality" for parabolic equations:

(43)
$$T(x_2, t) \geqq F\left(T(x_1, t)/B, |x_1 - x_2|/(t - t_0)^{\frac{1}{2}} \right),$$

provided $0 \leqq T \leqq B$ for $t \geqq t_0$. F is an a priori function, determined by c_1, c_2 and n. For the elliptic case where T is non-negative in a sphere of radius r centered at the origin, the result takes the form

$$(44)$$
$$\left|\log(T(x')/T(x))\right| \leqq H\left(r\left[r - \max(|x|, |x'|)\right]^{-1}, |x - x'|/r\right).$$

The a priori function H is determined by c_2/c_1 and n. This result is less easily obtained than (43).

Parabolic or elliptic problems with Neumann boundary conditions can apparently be handled by a relatively straightforward rederivation of the estimates of this paper in the context of the Neumann boundary, obtaining ultimately the Hölder continuity of the solution for any typical boundary shape.

References

1. L. Nirenberg, "Estimates and uniqueness of solutions of elliptic equations," *Communications on Pure and Applied Mathematics*, vol. 9 (1956), pp. 509–30.
2. L. Ahlfors, "On quasi-conformal mapping," *Journal d'Analyse Mathématique*, Jerusalem, vol. 4 (1954), pp. 1–58.
3. E. Rothe, "Über die Wärmeleitungsgleichung mit nichtkonstanten Koeffizienten in räumlichen Falle I, II," *Mathematische Annalen*, vol. 104 (1931), pp. 340–54, 354–62.
4. F. G. Dressel, "The fundamental solution of the parabolic equation," (also *ibid*, II), *Duke Mathematical Journal*, vol. 7 (1940), pp. 186–203; vol. 13 (1946), pp. 61–70.
5. O. A. Ladyzhenskaya, "On the uniqueness of the Cauchy problem for linear parabolic equations," *Matematicheskiĭ Sbornik*, vol. 27 (69), (1950), pp. 175–84.
6. F. E. Browder, "Parabolic systems of differential equations with time-dependent coefficients," *Proceedings of the National Acadamy of Sciences of the United States of America*, vol. 42 (1956), pp. 914–17.
7. S. D. Eidelman, "On fundamental solutions of parabolic systems," *Matematicheskiĭ Sbornik*, vol. 38 (80), (1956), pp. 51–92.
8. N. Wiener, "The dirichlet problem," *Journal of Mathematics and Physics*, vol. 3 (1924), pp. 127–46.
9. E. de Giorgi, "Sull'analiticità delle estremali degli integrali multipli," *Atti della Accademia Nazionale dei Lincei*, Ser. 8, vol. 20 (1956), pp. 438–41.
10. J. Nash, "The embedding problem for Riemannian manifolds," *Annals of Mathematics*, vol. 63 (1956), pp. 20–63.

11. J. Leray, "Sur le mouvement d'un liquide visqueux emplissant l'espace," *Acta Mathematica,* vol. 63 (1934), pp. 193–248.
12. C. B. Morrey, Jr., "On the derivation of the equations of hydrodynamics from statistical mechanics," *Communications on Pure and Applied Mathematics,* vol. 8 (1955), pp. 279–326.
13. J. Nash, "Results on continuation and uniqueness of fluid flow," *Bulletin of the American Mathematical Society,* vol. 60 (1954), p. 165.
14. J. Nash, "Parabolic equations," *Proceedings of the National Academy of Sciences of the United States of America,* vol. 43 (1957), pp. 754–58.

I was asked to contribute an "afterword" to this book and immediately a thought occurred to me in relation to the book and the context: the point of view of the person whose work and history become the subject matter of a book is different than that of the readers of the book. In a person's total life experience there is really no "inessential" and no "essential." The big thing is that a human has the opportunity and the experience of existence and life, and he or she may hope for reincarnation or to go to heaven when the life is indeed over and history.

It was also suggested that readers may be interested in what I have been recently doing of a scientific or academic character. Having been suddenly (in 1994) more highly recognized as a prominent contributor to what is called "game theory" than in years past, I began going to meetings and got into a research project which can be very simply described as concerned with the realization of "the Nash program" (making use of words made conventional by others that refer to suggestions I had originally made in my early works in game theory).

In this project a considerable quantity of work in the form of calculations has been done up to now. Much of the value of this work is in developing the methods by which tools like MATHEMATICA™ can be used with suitable special programming for the solution of problems by successive approximation methods. And a National Science Foundation grant was obtained to support a graduate student in the role of an assistant to help with this type of calculation and also to study the techniques of modeling contexts of bargaining and negotiation in terms of processes that can be investigated in terms of non-cooperative equilibria. (This part of the project work has not yet fully begun.)

I am working on a publication that will present the work on the project at its present state of progress, and also its projected continuation. But it seems inappropriate to go into much detail about the work and these plans in this afterword.

There are also a few ideas that relate to my older studies, from before the fall of 1994, which could conceivably, if not probably, develop into good research areas or achievements. So I have a few themes of broader diversification, with regard to research activities.

It is a phenomenon of these later times of my life that I have recently had frequent opportunities to travel, to go to meetings and the like. For example, I was able to see London and Berlin again, and had opportunities to see for the first time Athens and vicinity and Jerusalem. And I have had several opportunities for rather refreshing visits to various areas of Italy. I also made trips to various meetings in the United States of the sort that I had not been attending since the 1950s.

In 2002 I hope to visit China for the first time, and to see at least Qingdao and Beijing.

Sources

Chapter 1: Excerpted from a press release published by the Nobel Foundation, October 11, 1994.

Chapter 2: "John F. Nash, Jr.—Autobiography" was originally published in *Les Prix Nobel 1994*. Stockholm: Norstedts Tryckeri, 1995, pp. 275–79. Copyright © Les Prix Nobel 1994.

Chapter 3: "The Game of Hex" is excerpted from John Milnor, "A Nobel Prize for John Nash." *The Mathematical Intelligencer* 17, no. 3 (1995): 11–17. Reprinted with permission from Springer-Verlag. Copyright © 1995 Springer-Verlag, New York.

Chapter 4: "The Bargaining Problem" was originally published in *Econometrica* 18 (1950): 155–62. Reprinted by permission of the Editor.

Chapter 5: "Equilibrium Points in *n*-Person Games" was originally published in *Proceedings of the National Academy of Sciences* 36 (1950): 48–49. Reprinted by permission of the Author.

Chapter 6: Facsimile of John F. Nash, Jr., "Non-Cooperative Games." Ph.D. dissertation, Princeton University, 1950.

Chapter 7: "Non-Cooperative Games" was originally published in *Annals of Mathematics* 54 (1951): 286–95.

Chapter 8: "Two-Person Cooperative Games" was originally published in *Econometrica* 21 (1953): 128–40. Reprinted by permission of the Editor.

Chapter 9: Originally published as John F. Nash, "Parallel Control." RAND/RM-1361. Santa Monica, Cal.: RAND, 8-27-54. Copyright © 1954. Reprinted by permission.

Chapter 10: "Real Algebraic Manifolds" was originally published in *Annals of Mathematics* 56 (1952): 405–21.

Chapter 11: "The Imbedding Problem for Riemannian Manifolds" was originally published in *Annals of Mathematics* 63 (1956): 20–63.

Chapter 12: Originally published as John Nash, "Continuity of Solutions of Parabolic and Elliptic Equations." *American Journal of Mathematics* 80, no. 4 (1958): 931–54. © The Johns Hopkins University Press. Reprinted by permission of The Johns Hopkins University Press.

Photo Credits

Jacket photos of Nash: Courtesy of Martha Nash Legg and John D. Stier; 1, 11: Courtesy of Martha Nash Legg; 2, 4: Photos by Alan Richards, courtesy of the Archives of the Institute for Advanced Study; 3: Courtesy of Dorothy Morgenstern Thomas; half-title page, 5, 10: Courtesy Princeton University Library, University Archives, Department of Rare Books and Special Collections; 6: Photo by Herman Landshoff, courtesy of the Archives of the Institute for Advanced Study; 7: Brown University Archives; 8: Courtesy of Harold W. Kuhn; 9, 17, 18: Courtesy of John D. Stier; 12: Courtesy of Gertrude Moser; 13: Courtesy of the family of Ennio de Giorgi; 14, 16: Courtesy of Alicia Nash; 15: Robert Mottar, *Fortune;* 19, 20: Pressens Bild; title page, 21: Dick Pettersson, *Upsala Nya Tidning;* 22: Copyright © 2002 by Universal Studios. Courtesy of Universal Studios Publishing Rights, a Division of Universal Studios Licensing, Inc. All rights reserved; 23: Darryl McLeod; 24: Copyright © C. J. Mozzochi, Princeton, N.J.